Patton's Fighting Bridge Builders

Texas A&M University Military History Series

106

Joseph G. Dawson III, General Editor

Patton's Fighting Bridge Builders

COMPANY B, 1303RD ENGINEER GENERAL SERVICE REGIMENT

Edited by Joseph C. Fitzharris

Foreword by Earl E. Hall

Texas A&M University Press

College Station

First edition

The paper used in this book meets the minimum requirements of the American National Standard for Permanence of Paper for Printed Library Materials, z39.48-1984. Binding materials have been chosen for durability.

Library of Congress Cataloging-in-Publication Data

Patton's fighting bridge builders : Company B, 1303rd Engineer General Service Regiment / edited by Joseph C. Fitzharris ; foreword by Earl E. Hall. — 1st ed.

p. cm. — (Texas A & M University military history series ; 106)

Includes bibliographical references and index.

ISBN-13: 978-1-58544-550-9 (cloth : alk. paper)

1. United States. Army. Engineer General Services Regiment, 1303rd. Company B. 2. World War, 1939–1945—Regimental histories—United States. 3. World War, 1939–1945—Campaigns—Western Europe. 4. World War, 1939–1945—Engineering and construction. I. Fitzharris, Joseph C., 1946– II. Series.

D769.3351303rd .P38 2007

940.54'1273—dc22

2006011350

To America's fighting men and women, with gratitude and respect

Contents

Illustrations

Maps

Foreword

Joe Fitzharris has taken the complete diary of Company B, 1303rd Engineers and annotated and expanded it with information from U.S. Army archives in a fashion that each man of the company can appreciate. As an officer of the unit from its activation until it was effectively disbanded, I want to thank Joe for the excellent work he did. He got it right! This unit was made up of men from all walks of American life, but its core was made up of civilians with a background in heavy construction. I believe that with the bits and pieces of training we were able to provide during a helter-skelter year of organizing the regiment, these men made a contribution to the war effort that each can be proud of. Joe provides here a memorial my comrades and I can treasure. I thank him most sincerely for putting in the considerable effort to create it from the material available.

There is one man whose importance to the regiment and our war effort was very special. First Sgt. Robert Douds was the heart and soul of the unit. He was the best soldier I ever knew. I rode his coattails from platoon leader to battalion commander. I would like this book to go to press with full recognition of this outstanding citizen-soldier.

—EARL E. HALL

Preface

Company B and the 1303rd Engineer General Service Regiment (EGSR) were unique World War II units. EGSRs were supposed to function well behind the area of combat, yet the 1303rd (called the "Thirteen Third" by its soldiers) was in the combat zone from 28 July 1944 until the end of the war in Europe. Twice it guarded the right flank of Lt. Gen. George S. Patton's Third Army, first as it moved across France, then as it moved into Luxembourg and Belgium. Their service, and that of their sister EGSRs, the 1301st and the 1306th, in the Third Army, was so valuable and impressive in Europe that they were sent directly to the Pacific for the planned invasion of Japan.

For most of the war, the companies of the 1303rd operated independently, coming together in their battalions as needed. As a result each company has a separate and unusual story to tell. Company B was one of six "line," or operating, companies in the regiment; its unit diary forms the basis for this book. This story is not the regiment's story, but it gives the reader a good sense of engineer service in combat.

Capt. Earl E. Hall, Corps of Engineers, U.S. Army, owns the copy of the diary reproduced here. Hall led 1st Platoon, Company B, from the activation of the 1303rd at Camp Ellis on 15 July 1943 until he took command of the company in Luxembourg; in Japan he became commander of the first battalion of the regiment. He received this copy, a carbon copy (and not the first carbon either), before leaving Japan on 15 November 1945 to return to the United States for discharge. Hall thinks that the unknown author compiled the diary from the company's morning reports. Internal evidence suggests that the core compilation was done by clerks and then 1st Lt. Casey Deveikis (the company administrative officer) revised the Camp Ellis portion into a narrative. Since Deveikis left for the United States before the rest of the officers, this might explain the end of the narrative and the return to the daily report style.

What makes this company diary unique is that there are few engineer histories from World War II in print, and only one deals with general service regiments; none are company diaries. There were some published works on the 1303rd. Lieutenant Deveikis published four articles on the regiment in various issues of *Military Engi-*

neer, then wrote a history of the Thirteen Third. While he did a very nice job, his narrative attempts to tell seven separate (company) stories as one. Unfortunately, there are no source notes or attributions in the book, which is out of print. Worse, it was printed on poor-quality paper, and the available copies are not standing up to the ravages of time.

There are other works available on military engineers in World War II. The best source on the 1306th EGSR (a sister to the 1303rd) is Max Schwartz, *Bridges to Victory: Story of the 1306 Engineers in WWII* (Los Angeles: Schwartz Consulting Engineers, 2003). Dwight Gowdey wrote an article on the 1303rd entitled "Moving a Bridge 100 Miles" (*Military Engineer*, July 1944). Deveikis's articles include "The Eager Beaver" (*Military Engineer*, November 1947, 478–82) and "Grand Finale at Nihen-Seinen-Kan-Akasaka-Ku" (*Military Engineer*, January/February 1950, 48–51). The Reference Branch of the U.S. Army Military History Institute has compiled "Lorraine Campaign, Autumn 1944: A Working Bibliography," available online at http://www.ibiblio.org/pub/academic/history/marshall/military/mil_hist_inst/w/ww2eto5e.asc (the only item on an engineer unit referenced is Deveikis's history of the 1303rd). In addition to the unit diary, there is a typescript, "History of Company B," written by Tech 5 Robert S. H. McGlashan. Captain Hall owns an undated carbon copy (though it is from 1945). This document includes many fanciful events and alarums, invented dialogue, and other improbabilities. It probably is accurate enough on small details and on some of the "who did whats," but Hall recommended placing little faith in it. Comparison to the diary and to other sources does encourage a great deal of caution.

THE STORY BEGINS WITH the activation of the 1303rd EGSR at Camp Ellis, Illinois, on 15 July 1943 and ends at Tokyo, Japan, in November 1945 with the nearly complete replacement of Company B's World War II veterans by recent recruits. The compiler ends the diary on 13 November, just before he and Captain Hall left the regiment for discharge in the United States. The 1303rd continued in existence until deactivated in mid-1946. Its lineage continued through two short-lived units before being assigned, along with the regiment's honors, to the 130th Engineer Brigade (EB), activated at Hanau, West Germany, on 16 June 1969. Elements of the 130th EB saw Cold War service in West Germany, action in Operations Desert Shield and Desert Storm in 1990–91, and extended deployment in Croatia and Bosnia-Herzegovina (Operation Joint Endeavor) and during Operation Iraqi Freedom in 2003.

The editing process began with a photocopy of Captain Hall's carbon copy of the typewritten original diary. After scanning the document and correcting the resulting computer files, I identified acronyms, replacing nearly all of them with the full term referenced (for example, "CP" became "command post"). But what was a "S Sgt" (staff sergeant—and what was this grade?) or "RHQ" (regimental headquarters)? Such terms required explanations for the general reader as well as for the nonspecialist (who might know "S Sgt" but might not know what a "BB (D/S)" ["Bailey Bridge, a double/single" in this case] might be). The annotations are intended to help the reader without distracting from the diary's story. Additional material is interspersed throughout the text. Although the original diary was not divided in any way, I have here broken it up into logical chapters: Camp Ellis; Britain; service in the European Theater of Operations—Normandy-Brittany, movement east, Pompey-Toul-Nancy, Luxembourg, and Germany—and the Pacific Theater of Operations.

The text also needed context. It makes a difference to know that the company's first movement into the combat zone was along a single road from Coutances to Avranches. As the men moved with the rest of the Third Army along this one route, the Germans launched a counteroffensive to cut that road, but American units held their ground and the advance continued. Captain Hall remembers deploying with other engineers as infantry at about this time, though they never engaged the enemy. Similarly, knowing that, when the unit was on the right flank of the XII Corps (and thus the Third Army) with the Germans to the south, this was possible because the XIX Tactical Command (TAC) of the U.S. Army Air Force was actually protecting that flank so effectively that enemy ground forces involved mostly surrendered to the XIX TAC's commanding general.

Kerry Navarra was of great help as a reader throughout the editing process. He continually questioned and challenged me to make the diary clearer and more interesting for the general public. In addition, he scanned the bridge diagrams in the diary and Captain Hall's personal photographs. His careful work on the photographs was essential and greatly appreciated. For these reasons, he is a co-collaborator. Faith Bonitz and JoAnn Toussaint, the interlibrary loan people at the O'Shaughnessy-Frey Library, University of St. Thomas, were most helpful in acquiring the needed items from other institutions. Kenneth Schlessinger of the National Archives and Records Service, College Park, assisted me by locating the relevant regimental, XII Corps, Third Army, and SHAEF records. Christine C. Igielski, our administrative assistant, was—as usual—most helpful. My chair, Dr. Michael Jordan, and Dr. Winston Chrislock were both most supportive. Mark E. Jensen of the university's photo services did yeoman work on photographs and a map. Prof. Paul A. Lorah of the university's Department of Geography created the other maps.

Anne Looper Pryor of Washington, D.C., the daughter of Wilbur N. Looper, contributed letters and photographs of the company and her father, taken at Camp Ellis. Capt. Dwight Gowdey of Company D, 1303rd EGSR, also made photographs available. Kathy Robertson Hart provided copies of postcards from Rotenegg, Austria, where the company's last bridge was located. Doug Bekke, curator of the Minnesota Military Museum at Camp Ripley, provided the Cherbourg-Caen Sheet of the Nord du Guerre series used by Company B and other units in Europe. Donald Bittner, of the Marine Command and Staff College, was both encouraging and helpful, particularly in regard to the Articles of War. Capt. Earl Hall had the diary and photographs and spent a lot of time being interviewed. Earl was my friend long before he became the interviewee, and I thank him and his wife, Carol, for their friendship, cooperation, and support in this venture. My wife, Mary, was particularly patient in all this.

THE JOURNEY TO PUBLICATION began with Dr. James Sadkovich, acquisitions editor at Texas A&M University Press, who encouraged me to turn the diary into a publishable manuscript, then submit it to the press. I value Jim's friendship these many years, starting as it did on the northern plains. The internal and external readers provided thorough and critical commentary that produced a better manuscript. The members of the Texas A&M University Press editorial staff are extremely good at their jobs, most professional, and very delightful. I especially have appreciated the opportunity to work with Sarah Spring, Diana Vance, and Mary Lenn Dixon in the editorial and production stages. Kevin Brock's copyediting was superbly well done and greatly appreciated.

—JOSEPH C. FITZHARRIS

CHAPTER ONE

Introduction to Company B of the Thirteen Third

This is the diary of Company B, 1303rd Engineer General Service Regiment. It was activated at Camp Ellis, Illinois, on 15 July 1943. The 1303rd is unusual in being a supposedly noncombat unit that served in the combat zone with the Third Army for its entire time in Europe. The diary tells Company B's story from activation through its ending in occupied Japan. In England they built facilities for airborne and airborne "pathfinder" units, then the 1303rd deployed into France, where it served with the XII Corps, Third Army. During Patton's dash east in August 1944, and again in Luxembourg between the end of 1944 and February 1945, the regiment was effectively the flank guard of the Third Army. Finishing the European campaign at Passau, Germany, the 1303rd redeployed by fast ship through the Panama Canal to the Philippines to prepare for the planned landings on Kyushu in November 1945. Following Japan's surrender, the regiment became the engineer unit for general headquarters in Tokyo. The diary ends when the owner, Capt. Earl Hall, was ordered to the States after most of the original members were already en route home.

Engineer General Service Regiments were intended to be construction units that would build and repair airstrips, hospitals, buildings, prisoner-of-war enclosures, roads, bridges, and such behind the combat front in the "communications zone" (or "COM-Z"), where they would not come under attack. Many of the EGSRs became little more than organized labor pools, particularly the segregated black units. The 1303rd, called the "Thirteen Third" by its men, was different. From 28 July 1944 until the end of the war in Europe on 8 May 1945, they were continually in the combat zone, removing mines and demolitions, repairing or building bridges, repairing roads and buildings, serving as the flank guard for the Third Army in August 1944 and again in December–January 1945, and twice deploying as infantry to help defend against a German attack. Using engineer units as flank guards was unusual enough, but employing a GSR was nearly foolhardy on the face of it. Company B was in the thick of this activity, and its company diary helps tell this tale in some detail, providing the opportunity to look below the narrative history of a regiment at some of the elements that make up that story.

This copy of the company diary was given to Captain Hall shortly before he left the unit in November 1945, bound for the United States. Hall commanded the 1st Platoon from activation until 16 January 1945, when he became company commander. On 8 October he was given command of the 1st Battalion. The material in the diary clearly comes from company morning reports. Someone compiled this information into a diary, but who he was is unknown. The accuracy of the account is supported by the regimental S-3 journal in the National Archives at College Park, Maryland, and by 1st Lt. Casey Deveikis's history of the regiment.[1] The first portion, detailing the unit's time at Camp Ellis, was converted into a narrative, probably by Lieutenant Deveikis, then the executive officer of the company. The diary is unusual in offering access to daily events nearly as they happened, with minimal editorial additions.

Although Hall only commanded the company for about nine months, he had more influence on the unit than the time suggests. We can see in the diary that Hall was the "go to" officer, always either attending schools, serving on detached duty, or on other special assignments. It is also clear that the men of his platoon, then the company, came to have a bond with him, and he with them. (The captain was able to provide explanations and commentary as part of this publication.) Hall identified 1st Sgt. Robert L. Douds as the key man in the company. Officers have their roles, and the first sergeant has his. Douds, as all good first sergeants should, knew his role and place, keeping the men functioning and behaved. The story of Company B might well be the story of Robert Douds. It could be the story of Earl Hall too. It is certainly the story of Wilbur Looper, Logan Stoudt, and all the other men who became part of the unit at Camp Ellis, Illinois, in the summer of 1943 and served throughout the cold winter of 1944–45 in Luxembourg. It is the story of the men who joined at the Calas staging area in Marseilles, France, and went on to Japan in September 1945, thankful for not having to fight their way ashore in the Operation Olympic landings that had been planned for that November. But the focus of this book is the company: what it did, when and where, and who did it.

World War II was, to put it tritely, a global watershed event of incalculable proportions. And it forms both the origins and the context for this story of honorable service. For our purposes, these beginnings were the national defense buildup in 1940 after the German conquest of France and the Low Countries, followed by the introduction of conscription on 16 September. In May 1941 Pres. Franklin Roosevelt declared an "unlimited national emergency." Somewhere during this period, if not earlier, Douds and a number of other senior sergeants entered service. With the emergency, the typical one-year term of conscript service was changed to "the duration plus six months."[2]

Earl Hall (class of 1943) and Wyman Moseley (class of 1944, though he did not graduate that year) were in college—at Texas Agricultural and Mechanical Institute. The school required all students (who were all men) to serve two years in the Reserve Officer Training Corps (ROTC). Before 1942 Hall was in the enlisted reserves, and that summer, when ROTC summer camp was cancelled (for all units), he went to school. Like many other colleges, Texas A&M offered accelerated coursework so students could graduate before entering service. In January 1943 Hall graduated from Texas A&M. Because he and his classmates had not completed summer camp, though, they were not eligible for officer commissions. They had three options instead: become a civilian and await the draft, attend Officer Candidate School as a civilian and either earn a commission or fail and return home to await the draft, or enlist immediately as a corporal and go to OCS and either get a commission or fail and remain a corporal. Like many others, Hall, and later Moseley, chose corporal's stripes.[3]

By the spring of 1943, it was clear that the United States would need a larger number of engineer units, both combat and general service, than had originally been thought. The lessons from the North Africa campaign, America's first in the European war, were hard bought and, perhaps because of that, responded to with alacrity if not acuity.[4] Army Ground Forces command sought to change general-service regiments into battalions and to strip them of much of their equipment. (On 9 March 1942 structure of the U.S. Army was reorganized into three operational components: the Army Air Forces, Army Ground Forces, and Army Service Forces.) Army Service Forces command and the Office of the Chief Engineer, head of the Corps of Engineers, opposed this. The end result was a reduction of these regiments in both size and capability. (See appendix 1 for further discussion.)

This need for more engineer units was paralleled by the need for more medical, ordnance, transport, quartermaster, signals, and other support units. But existing training facilities were inadequate to meet this need. In response, Army Service Forces opened two new unit-training camps, Camp Claiborne, Louisiana, and Camp Ellis, Illinois. The Corps of Engineers activated training command groups at these camps. Fresh from OCS and engineer-officer training at Camp Belvoir (now Fort Belvoir) outside Washington, D.C. (near Mount Vernon), Hall and Guy Bubb, a Texas A&M classmate who served in Company A, reported to Camp Claiborne and after a few weeks were sent as cadre officers to Camp Ellis to form the 1303rd EGSR. At the same time, Douds and a number of noncommissioned officers (NCOs) and enlisted men also arrived at Camp Ellis.[5]

CHAPTER TWO

Forming the Unit at Camp Ellis

JULY 1943–MARCH 1944

Camp Ellis had its beginnings in September 1942, when the army purchased over seventeen thousand acres of relatively flat land in Fulton County, southwest of Peoria, in west central Illinois and began demolishing farmsteads. Within seven months the first servicemen arrived at the new unit-training center for Army Service Forces. In time, Ellis came to have more than two thousand buildings, railroad sidings, an airstrip, five thousand German prisoners of war, and as many as 40,000 U.S. troops. Over two years 125,000 men in 456 units (ranging from 5 to 1,200 men in size) received training there. When the 1303rd was activated, the Camp Ellis was newly opened and its facilities were rudimentary. Constructing the other facilities was an intentional part of the unit's training program.[1]

The initial buildings were wood framed and sheathed with tarpaper siding and roofing; the addition of siding came later. The interiors were raw lumber, though postcards show finished walls and ceilings.[2] Service clubs, chapels, theaters, barbershops, a shoe-repair shop, rifle ranges, the motor pool, and many other buildings and fixtures for various types of training were built. Ultimately Camp Ellis contained one of the largest army hospitals, both to serve as a training facility and to handle wounded servicemen returned to the United States.

When the cadre of the 1303rd EGSR reported to Camp Ellis in July, they began a training cycle that featured requalifying with the rifle, building and then using a series of obstacle courses as they moved from point to point about the camp, and taking a multiday hike. The turmoil in the ASF and the Corps of Engineers over what trainees needed to know and how long it would take them to learn it, both as individuals and as a unit, are reflected in the Thirteen Third's experiences at Ellis. While they struggled through the Illinois mud and snow, the Allies landed first in Sicily, then on the boot of Italy, and began moving northward. The Allied air forces bombed Germany daily. In the Pacific the U.S. advance through the Solomon Islands and up New Guinea were both well underway. The submarine campaign against the Japanese began to show signs of success. And while the German submarine campaign against shipping raged vigorously, by the time the regiment shipped out for England, the Allied navies had turned the corner in the crucial Battle of the Atlantic.

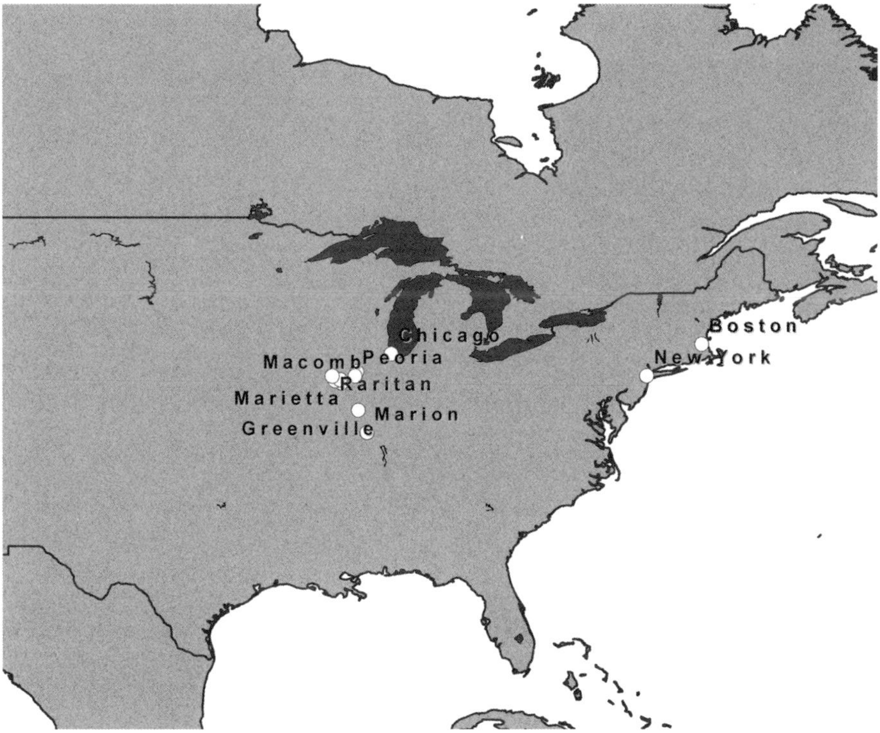

The Eastern United States. Courtesy Paul A. Lorah

The Diary of Company B

USA

CAMP ELLIS, ILLINOIS

ACTIVATED 0001[3] 15 JUL 43—DEPARTED 1530 9 MAR 44.

THURSDAY—15 JUL 43

Company "B" and the rest of the 1303d Engineer General Service Regiment[4] was activated at 0001 hours on Thursday, 15 July 1943,[5] at Camp Ellis, Illinois, the Army Service Forces Unit Training Center located near the town of Peoria. The Company Commander was 1st Lt Manuel F Gonzalez.[6] His outfit consisted of one officer, 2d Lt Earl E Hall,[7] and 13 enlisted men,[8] the cadre[9]: Sgt [Robert L.] Douds,[10] Cpls [Howard P.] Cooke, [Joseph D.] Sheridan, and [Clifford M.] Slade, Tecs 5 [Robert I.] Hawke, [George] Leder, [John C.] Olson, and [Joseph J.] Willmeth, Pfcs [George] McMahon and [Richard E.] Murphy,[11] and Pvts [Roger G.] Dunbar, [Melvin G.] Jenkins, and [Robert E.] Pennington. The Regimental Commander was Lt Col

Cadre, 1303rd EGSR, Camp Ellis, 1943. Major Vandenburg is standing to the right; Private Looper is in the top row, to the right side of the sign. Courtesy Ann Looper Pryor

Richard W Pearson. [Informal] activation occurred . . . when Lt Gonzalez introduced himself to Sgt Douds as the Company Commander and the Sergeant presented himself as his First Sergeant. This transpired on the rifle range. It was their first meeting.

Records of the first days are unavailable. That at first flush seems unfortunate. It is not. If the records were available, that would be unfortunate. The things that happened during the approximate period 15 July–1 December are better left unsaid in writing. Suffice it to say, utter chaos reigned worse confounded. Illustrative of what must have happened in those early days is the Camp Ellis Classification system. The enlisted cadre for the Regiment was formed according to height, single file. The Company cadres were formed by dividing this file of men into seven parts, the tallest becoming Company "A," the next tallest Company "B," and so on down the line. This to form a unit of specialists!

AN ENGINEER GENERAL SERVICE REGIMENT (EGSR), like all other noncombatant Army Service Forces (ASF) units, operated in the "communications zone." They "were more concerned with building shelters, roads, ports, or air fields, and with performing supply functions." Combat engineer units were "to clear and construct obstacles, lay mine fields, ferry troops in river crossings, build bridges and, . . . act as infantry." After some experience in North Africa, Army Ground Forces sought to have EGSRs reorganized into battalions. The ASF disagreed, and the end result was that EGSRs retained their regimental structure but were reduced to six line companies organized in two battalions. Ultimately, twenty-nine white and fifty black EGSRs were organized. Only three (the 1301st, 1303rd, and 1306th) EGSRs served in the combat zone, all in the Third Army under Lt. Gen. George S. Patton and only because of a shortage of combat engineers.[12]

Whatever else may be said for the Regiment, it got off to a running start with both feet. Four days before activation, and one day after arrival on the post, cadre officers and men were starting a ten-day period of rifle marksmanship instruction.[13] After that everyone arriving at Camp Ellis spent the first ten days on the rifle range: everyone. An "Expert" badge earned the week before at Fort Belvoir, for example, did not count at Camp Ellis. Activation was only an added ingredient on a single day during the strenuous course.

This training completed, a short course of instruction for the cadre was instituted. Unfortunately, it did not last beyond a couple of days. The training facilities of the post had to be developed. This task fell to the lot of the 1303d. Obstacle courses were started, obstacles built, roads cut where no roads were before, buildings built. On 23 July the initial consignment of fillers arrived, 57 of them for Company "B." They came from the 361st Engineer Special Service Regiment, Camp Claiborne, Louisiana. The new arrivals were, of course, taken out to the rifle range. Part of the cadre acted as instructors; the remainder carried on with the construction of the camp.

Throughout the month of August new men were added to the Regiment, and then processed out as quickly as possible because a great many of them were physically unfit not alone for an Engineer organization but even for the military. No official confirmation of this has ever been obtained, but opinion in Ellis ran high that the 6th Service Command was screening itself out and dumping the least qualified and capable men into the new units whose activation and training was being rushed by the War Department but for which there were not adequate personnel.[14] The Company in about three months processed in and out again at least three hundred men. And this, for the most part, during basic training. The 1303d was getting ready for war!

On 6 September 1943 the Regiment was relieved of responsibility for completing the training facilities and ordered to begin basic training under MTP 5–1.[15] It was not a very successful training program, that which we followed. The allotments of time were inadequate. The procedures were

Private Looper with truck, Camp Ellis. Courtesy Ann Looper Pryor

not accurate and in many instances were being taught to people who knew nothing about them by people who knew only a little more. Too much time was spent compiling a myriad [of] reports and manufacturing what is known as "eyewash." And on top of it all, people were going and coming like mad. It was a time of what is known as a "rat race." One of the coherent things accomplished during this time was in bridging. Under Lt Walter Solomekin, Company "D," our Sgts Faucett and Sheridan were part of the crew that removed the Chillicothe steel bridge, a 210-foot span, floated it down the river on barges, took it apart, transported it to Camp Ellis, and there re-erected it for [the] training use of our own and the other regiments of the Engineer Group.[16]

The Regimental Command changed early. While Major Vincent I Vanderburg [*sic*],[17] the cadre Executive Officer, was heading a bond-selling entourage of men and equipment from the 1303d between 4–19 September,

Lt Col Pearson was relieved of command. This occurred on 8 September. Major Wallace V Fretts, senior officer present, assumed command. On the 20th, Major Fretts was relieved and Major Vanderburg, one day returned from selling bonds, became the Regimental Commander.

As part of the training program Lts [Casey] Deveikis[18] (who reported to the Regiment for duty 8 Sep and was assigned to the Company on the 9th), Hall, and [Arnold] Saggese (who joined the Company very early in August) took 12 enlisted men [abbreviated as EM] to Fort Sheridan, Illinois, for a three-day training course in the 50 calibre machine gun school there. They were part of the large Regimental detachment ordered there between 27–31 October. Major [Neil S.] Kingsley commanded this group while on DS [detached service].[19]

Training continued apace. Air strips came next in our unusual program: a technical subject, part of a later program, but before we were done with basic subjects. There was the whole rub. Camp Ellis was a unit training center, but the people there had not had their individual training.[20] Both activities were carried on *simultaneously* with the tacit understanding that this would never be admitted and thus eventually it would work out for the best. It did not.

Air strips, then. The fourth 2d Lt, Weyman M Moseley, Jr,[21] arrived on the 6th of November, in time to change from pinks to fatigues and go out to the air strip which had just been completed the night before after more than 30 hours of work in filthy weather and finish the trimmings on the job, clinching the mats. It was not until he was done with his first assignment that he had a chance to report to his Company Commander. The arrival of the lieutenant and his platoon was almost simultaneous. A fresh group of inductees came to us at that time, and they were assigned to the new 3d Platoon commanded by Lt Moseley.

Three days after his arrival (Lt Moseley's), Lt Saggese left the Company, transferred to the Military Censorship School.

Big things were in the offing. On 12 November a practice march was undertaken between the hours 1300–2245. It should have ended sooner, being only 20 miles, but the guide got lost so we walked about 32 miles. Little did we dream that 32 miles would be small potatoes ten days later.

THE HIKE

STARTED 1200 18 NOV 43—ENDED 2400 21 NOV 43.

The long walk across Illinois is for us a major catastrophe not unlike the Flood sent to castigate the sinful of those days. At high noon of Thursday,

Assembling a foot bridge during training, Camp Ellis. Private Chalom is standing to the right. Courtesy Earl E. Hall

Setting a bridge abutment during training, Camp Ellis. Courtesy Earl E. Hall

18 November, three officers and 168 enlisted men took off on a walk, passed through Table Grove at 1400, halted twice—at 1630 and at 1740—and bivouacked at 2140, 21.5 miles from RHQ [regimental headquarters], near the farm of one E. E. Allison. The bivouac party, consisting of mess personnel and other men assigned camp duties, traveled by truck ahead of the marching Regiment, selected and prepared the sites, and cooked the meals. We only walked.

The second day started at 0600: that is, the walking started then. We passed through Macomb at 0730, halted between 1000 and 1100, went through the town of Blandinsville at 1515, and bivouacked ten minutes later about a half-mile north of that town. We had come 42.5 miles.

On the 20th, the third day of the hike, we started trudging at 0300. At 0715 we passed through Raritan, halted from 1010 to 1100, went through Youngstown at 1425, and settled down for the night at 1630 southeast of Walnut Grove. The mile-o-meter then stood at 72.2 miles. Our feet hurt.

Came the fourth day, the 21st. Again it started at 0300. By 1100 we were at Marietta. At 1215 we bivouacked about 2.5 miles east of town. Resuming the

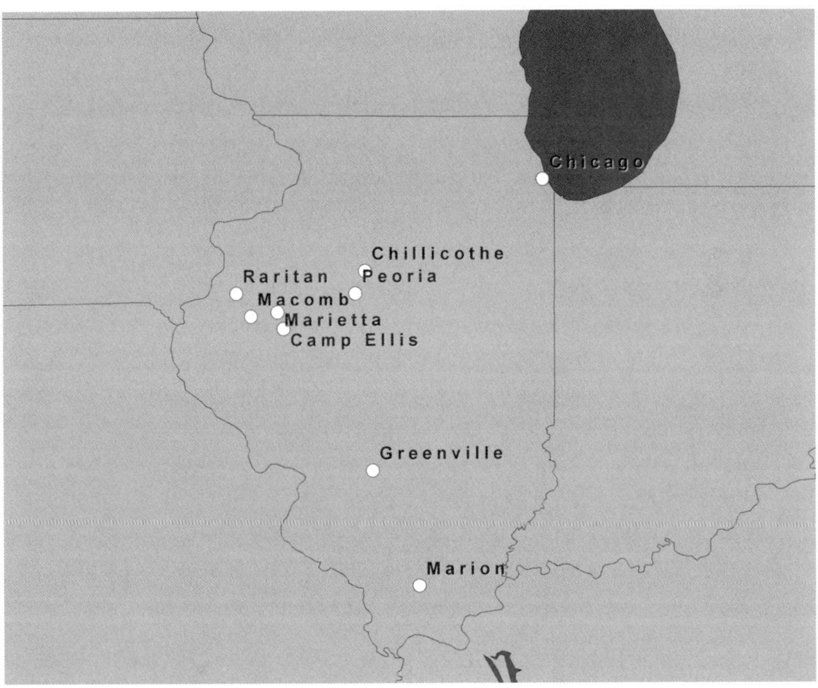

Illinois. Courtesy Paul A. Lorah

hike at 1800, we halted for thirty minutes at 2100 south of Smithfield. That was the last break, for at 2400, after climbing the long hill from the rigging area (an impossible feat until it was done), we were back at RHQ whence we had come some days earlier when we still had legs to walk on instead of bloody stumps. We had walked 108 miles in three and a half days and that night were undoubtedly the weariest soldiers in the American Army. The next afternoon, when we had to fallout as a Regiment on the edge of camp, and stage a triumphant return from our hike in order to breast [*sic*, to break] the finishing tape to the ovations of the Engineer Group, we were undoubtedly the angriest soldiers in the American Army.

We all talked about it for days afterwards, but never did we arrive at a conclusion, what had been proved. There could be no question that the hike was valuable training then, training for war. Then we did not need to know that through two wars we would never walk more than three city blocks at any given time. When it was all over we got certificates with many mimeographed signatures on them. The people who had signed had driven out in their sedans to see us march when we first started the hike. Lt Hall, unfortunately, missed a singular experience when he was detailed to the Second Army Mine School at Camp Forrest, Tennessee.[22] He left the day we did and returned on 1 December. He does not have a certificate for the 108 mile march.

Following the hike three day passes were given the entire command. Thanksgiving Day came on the fourth day which, for us, was a day of: work. The evening meal was a very real compensation. Mess Sergeant Dunbar's kitchen did themselves proud in turning out the magnificent supper which was the first holiday meal for Company B, the first of what was to become a longish series of them.

A wicked personnel problem was contained in the phrase "over strength in grade."[23] Not until late in the history of the Company — December, January and February — were we able to promote our noncommissioned officers to their proper grades. On 3 December Lt Hall received his promotion to full lieutenancy. On 1 Jan 44 Lt Gonzalez was promoted to Captain.

Basic training, according to the schedule, ended 3 Jan 44. Inadequate as it had been, it was the basis upon which the Regiment and the Company was founded. Following this period of intensive work, the entire command was granted leaves and furloughs. While groups were changing, some attempt was made to compensate for the shortcomings of the training program. Not a great deal was accomplished. Finally, as the climax of training, the Regiment was ordered to conduct its own field exercises for two weeks early in February.

MANEUVERS

STARTED 0830 2 FEB 44—ENDED 0130 12 FEB 44.

The Regimental Maneuvers we remember for the ragged country in which they were conducted and for the icy weather in which we nearly froze returning to Camp Ellis. We departed the Camp in trucks on Wednesday morning, 2 Feb 44, got to Greenville and one mile south of it by 1730 and there set up an overnight camp. The next morning we started at 0830, arrived at Shawnee National Park by 1700, and established camp, some 94 miles from our area at Ellis and 15 miles south of Murphysboro, in the vicinity of Pomona.

Maneuvers began the next day.[24] We performed an Engineer Reconnaissance and made a tactical bivouac.[25] Nobody explained in any critique why an Engineer Reconnaissance was being performed in rocky hills two and three hundred feet high with steep bluffs on every side, but that did not stop us from learning a few fundamentals of mountain climbing. We liked it so well, we repeated the performance on the 5th.

The Company moved into a defensive position on the east side of Muddy River on the 6th, worked on combat principles the day after that, and on the following day—the 8th—"B" and "C" Companies attacked Company

Machine-gun emplacement during maneuvers, 1944. Courtesy Earl E. Hall

Moving heavy aluminum ponton during training, Camp Ellis. Courtesy Earl E. Hall

Positioning heavy pontons for bridging, using an outboard motor for power during training, Camp Ellis, 1943. Courtesy Earl E. Hall

Launching heavy rubber pontons for bridging during training, Camp Ellis, 1943. Courtesy Earl E. Hall

"A" following a tactical move into position.[26] Late that night we moved up for camping to the top of Bald Knob, the coldest and windiest hilltop in the state of Illinois. On the 9th we rested. The final day of maneuvers was spent nontactically, building bridges at the CCC Camp there, Camp Pomona.[27]

The two-week schedule was called off upon receipt of Movement Orders from the War Department. We departed Camp Pomona at 0830 on the 11th in extremely cold weather. We halted for dinner at Greenville and proceeded to Camp Ellis, arriving there at 0130 of the 12th of February, Saturday morning. Although equipped with Arctic clothing for the field exercises, we found it inadequate for warmth on that return trip in the snow and ice. To a man, we were glad to see an end of maneuvers and of all training. After a hot shower and a chance to thaw out in bed until nearly noon, we began thinking of things to come as we started preparing for overseas shipment.

Fifth officers[28] are something we have not fared well with in Company "B." On 14 Feb 1st Lt George B Richards, one time Commanding Officer of Company "D," was assigned to us. Thereupon officer assignments were changed in the Company. The new lieutenant replaced Lt Deveikis as the Company Executive Officer who assumed command of the 2d Platoon, which he had commanded along with Company Headquarters during the hike and the maneuvers. This arrangement did not last long as Lt Richards was detailed as a member of the secret advance party for the Regiment and departed the Company on the 22d of February. He was formally relieved of assignment on the 25th of that month, whereupon Lt Deveikis left S Sgt *Cooke*[29] in charge of the 2d Platoon and returned to his duties as Administrative Officer and the hectic business of getting the Company ready to ship. Lt [Donald] Geesey was assigned to and joined the Company 20 Feb 44.

The authorized strength of the company, or any "line" company in an EGSR, was five officers and 168 enlisted men.

The POM [Preparation for Overseas Movement][30] was long, confused, and unnecessarily difficult, as first things are bound to be. Equipment was checked for condition before being processed for shipment and repaired or salvaged. Clothing showdowns were held and deficiencies corrected. "Shots" were administered against virtually every known disease in Europe. And in the midst of all these furious preparations, there were heard the voices of the doubting Thomases: "This outfit will never get past a P of E [Port of Embarkation — sometimes POE]."[31] Some of these speakers really believed what they said.

ENROUTE.

WEDNESDAY — 8 MAR 44 TO FRIDAY — 10 MAR 44

The final Morning Report at Camp Ellis was submitted for Wednesday, 8 March 1944, the date of departure from our home station. We boarded the train at Ellis, our destination unknown, and proceeded on an uneventful ride to the staging area.

CAMP MYLES STANDISH

ARRIVED 0330 11 MAR 44 — DEPARTED 1110 21 MAR 44

SATURDAY — 11 MAR 44 TO MONDAY — 20 MAR 44

We arrived at 0330 in the morning, a bad time for any arrival anywhere. . . . The closest town was Taunton, Boston was farther away. . . . Here POM requirements were completed in confusion and eventually we got a ship assignment.

THE POE FOR BOSTON was Camp Myles Standish. Some units were there long enough to obtain passes into Boston, but it is unlikely the men of the Thirteen Third were given much chance to gawk at the city's many delights. They were assigned to the SS *Argentina*, a former passenger ship taken up by the government for service transporting troops.

CHAPTER THREE

England

MARCH–JULY 1944

In March 1944, as the SS *Argentina* and its convoy crossed the Atlantic, the crews of the escorting ships worried over their charges, moving no faster than the slowest merchantman, and searched anxiously for lurking U-boats. The Battle of the Atlantic had reached its worst point a year earlier, but German submarines were still a menace. The soldiers, at least those not below decks heaving up their breakfasts, likely found the movements of the escorts inscrutable; only once does the company diary note the dropping of depth charges.[1]

Landing, as did many troop convoys, at Gourock, Scotland, at the mouth of the River Clyde, the soldiers disembarked and boarded trains for the trip south. The carriages were uncomfortable, unheated, and overcrowded. In addition, the newly arrived Americans had their first taste of coffee as the English make it. The thirteen-hour trip to the town of Wellingborough ended with a two-hour truck ride to U.S. Army Air Force (USAAF) Station 541 at Riseley. There the 1303rd became part of the engineer force building bases to support the invasion of Europe, assigned to District VIII Eastern Base Section.[2] The men were sent out to prepare facilities for the airborne units, including an airborne pathfinder unit, at various USAAF stations. Once the invasion of France, Operation Overlord, was underway, the regiment concentrated at Riseley and awaited orders to land on the Continent.

The Germans considered Lt. Gen. George S. Patton Jr. to be the best American combat commander. But because he had struck two enlisted men hospitalized for combat fatigue during Operation Husky in Sicily, Patton was initially "shelved" pending further disciplinary action. Later, realizing that Patton's talents would be useful, Gen. Dwight Eisenhower gave him both a future — command of the U.S. Third Army after the invasion force gained a secure foothold — and a more public nonoperational command, the "First U.S. Army Group." This "army," in reality merely a large number of inflated balloons made to look like tanks, trucks, and buildings along with a number of signals personnel sending radio messages from one nonexistent unit to another, was part of a deception scheme called Operation Fortitude.

Identifying Calais as the best landing site along the English Channel, the Germans expected that the Allies would be as logical as they in determining where to put troops ashore. Instead, Eisenhower chose to land at what the high command considered inferior locations, hoping to achieve defensible beachheads or even break through into the interior before the enemy could bring up strong reinforcements. Patton's "army group" was situated opposite Calais, which with his reputation, helped focus German attention on that region for weeks while preparations for the actual invasion proceeded.[3]

After Allied forces had secured their position in France, Patton prepared to take command of the Third Army upon its activation on 1 August 1944. In July he shifted to the Normandy area and assumed operational control of VIII Corps (to which the 1303rd would first be attached) and some other units of the forthcoming Third Army. A cavalryman who shifted to armor and also earned his pilot's license, Patton's way of war was very different from that of Lt. Gen. Omar Bradley, commander of the (real) Twelfth Army Group, or Field Marshal Bernard Montgomery, the British commander of Allied ground forces and the XXI Army Group. He applied cavalry speed and dash to his operations, while infantrymen Bradley and Montgomery approached war more methodically.

Meanwhile, the Thirteen Third waited in England.

England, March–July 1943

AT SEA

ABOARD USS ARGENTINA

EMBARKED 1500 21 MAR 44—DEBARKED 1520 4 APR 44.

TUESDAY—21 MAR 44 TO THURSDAY—23 MAR 44

The Regiment was detailed as the advance party for the shipment and boarded the "Argentina," a converted passenger liner on the South American run, at 1500 in the afternoon. Our boat order was 589. Aboard the ship after the men were berthed down, the work parties began setting up shop to function in the galley, as guards, and in other work assignments.

FRIDAY—24 MAR 44 TO SUNDAY—2 APR 44

We sailed from Boston Harbor at 0120–0210: the time interval between hoisting anchor and actually steaming off under our own power. The fourteen day crossing of the Atlantic was without incident. Once some depth charges were dropped towards the end of our voyage. We traveled in convoy under strict blackout at night, and everyone was unnecessarily a little wary on this initial crossing. There were Army nurses aboard the ship. The food was poor and the men complained of it with justification. At night the darkened ship was hot and stuffy, and at the end it was good to get off and on land again.[4]

SCOTLAND

ENROUTE

MONDAY — 3 APR 44 TO TUESDAY — 4 APR 44

After sighting the coast of Scotland we came into the harbor of Gourock through the Firth of Clyde and settled down to spend the night at anchor.

Scotland was good to see. What little we saw of the countryside prior to debarkation — which took place by tender from the anchored ship at Gourock to Greenock and thence onto land at 1520 — Tuesday, the 4th of April, 1944 — was a most pleasant aspect. It looked, all of it, like a well kept garden, neat and green and trim in the early spring.

Upon debarking we had time to wait a few minutes for the train the boarding of which — at 1615 — was made more pleasant by doughnuts and horrible coffee provided by the British Red Cross.

ENGLAND

RISELEY, BEDS — 1–1/2 MI N, WL4984, ENG GRID[5]

ARRIVED 0700 5 APR 44 — DEPARTED 1020 17 JUL 44.

WEDNESDAY — 2 APR 44

The train ride through Scotland lasted all night. At 0500 in the morning we arrived at Wellingborough where we detrained, had breakfast, and then mounted our trucks for the ride to our permanent station in the United Kingdom. We arrived at 0700 hours at Melchbourne Park which was also known as [US]AAF[6] Station 541[7] or, technically, by the grid coordinates listed above. Everyone went immediately to bed, the English day coaches not having been pleasant or comfortable, and it was not until the afternoon that people began stirring to the start of their war work.

The advance party had just completed setting up our tent city before we arrived. It was next door to the largest mustard gas generating plant in the UK, in the middle of an unbelievably large bomb storage depot.[8]

THURSDAY — 6 APR 44 TO WEDNESDAY — 26 APR 44

. . . The first days were filled with housekeeping and with the getting of our equipment, always a complicated process. The area in which we lived had to be improved. When the roads and sidewalks were concreted, the order of all orders was issued: there is no mud in England. But there was mud, great gobby seas of mud which were forever being tracked up on the nice new concrete. The inspecting officers from the Com Z [Communications Zone][9] installations of now Lt Gen Lee and now Brig Gen Plank were as strict as their masters. We had been assigned to District VIII, Eastern Base Section to work essentially in completing the building of the English base which would be used as the springboard for the invasion of Europe. We worked cutting out roads, pouring concrete, building hard standings for bomb storage. The pace was terrific, garrison life was stressed and garrison conditions required in the field and these conditions obtained after a while.

Then came an interesting assignment: the Regiment was given the task of establishing the camps on various airdromes for the use of airborne troops who would spearhead the invasion. Five stations were assigned to Company "B."

The assignment was "Top Secret." The Colonel pooped up the officers the night before we left, but there was only a little he could tell us about the job: only that we would have something to do with the paratroopers who were to spear-head the invasion of Europe.

THURSDAY — 27 APR 44

Early in the morning we began leaving Riseley for our various stations, all AAF fields. This departure could not [be] obvious, so we left at thirty minute intervals. Captain Gonzalez took the 1st Platoon to Langar, Station 512.[10] Lt Deveikis went to Spanhoe, AAF field 493 near Stamford with 13 of his men. Lt Moseley took his squad to Bottesford (in Lincolnshire), Station 481. Lt Geesey with his detachment departed for Fulbeck, AAF Station (488 in Lincolnshire). Capt [Morris S.] Bolzer, 1st Battalion[11] Adjutant, took Sgt [George] Rohrmann's squad to Station 482, Balderton (Nottinghamshire). This was done because Lt Hall was absent at Mine School near Oxford.[12]

FRIDAY — 28 APR 44 TO MONDAY — 5 JUN 44

At [the] same time we departed Company "A" sent out its four detachments to the remaining stations assigned to the Regiment. These were located at Salisbury, Barston Heath, Folkenham, and Cottesmore.

After several weeks Lt Moseley was transferred to North Witham, AAF Station 479, where he established the field for the Pathfinders — the trail blazers for two Divisions, both Airborne: the 82d and the 101st. Thus it was we found out at our stations that we were working for the IX Troop Carrier Command. Our job was to set up housekeeping for these two Divisions of airborne troops at the fields where they would stage for briefing prior to departure to start the invasion of Europe. Accordingly, we began to erect tent cities near hangars designated to house the troops. We set up messes and established latrine facilities (and, incidentally, learned all about honey buckets[13]) and provided CPs and briefing rooms with sand tables for terrain studies. On each of the fields these special areas were enclosed behind

Scotland and England. Courtesy Paul A. Lorah

barbed wire to keep the briefed paratroopers from contact with unauthorized personnel.

The balance of the Company, meanwhile, was fighting the war with mud and impossible assignments at Riseley. The 1st Sgt was acting Company Commander, an experience he never tires of relating. "Hectic" is the word for what he went through. Messes were consolidated and ours began feeding an unknown number of men—from 30 to 300—which could (and did) vary on a minute's notice; sometimes, even without notice. The concrete driveways and sidewalks had to be practically polished to keep the mud off, and mud was being tracked on consistently. Air compressors were resorted to as a last—and successful—measure. Everything was painted olive drab: everything that could be painted, that is. That excluded only the canvas under which we were living and the concrete in the area. The labor bore fruit. From time to time the detachments sent couriers back to Riseley for mail before their stations were sealed up upon arrival of the paratroopers, and these couriers, after one breathless and awe-struck look at the camp they had left, pronounced it beautiful. And it was: nowhere could a more model example of a garrison establishment have been found in the field. Contradictory, perhaps, this was: it produced "superior" ratings when the Communications Zone brass hats inspected the 1303d Engineers, though, and that was what we were after. The wailing of tears and the gnashing of teeth that occurred to get these results was the lot of 1st Sgt Douds who has never got over the things he had to do. Certainly he is the most unenvied man in the Company.

Finally, on Monday, the 5th of June, 1944, the good word came. The planes were loaded. The paratroopers donned their invasion regalia. After supper they boarded their planes, these grinning, joking, impious paratroopers, perhaps realizing that more than half of them were going to meet death, perhaps not thinking about it. The planes of the main body of both divisions took off between the hours 2200–2400. On Lt Moseley's field at North Witham, though, the Pathfinders had gone earlier that evening. They dropped into enemy held territory about 12 hours before the invasion began to set up and mark the drop zones for the airborne army that was coming early the next morning.

Lt Hall had returned to the Company about a week after our detachments were gone. He remained in command about a week before he was sent out on detached service also: to Depot G-15 where he worked on invasion maps under heavy guard, breaking them down for using organizations and distributing them.[14]

Lieutenant Hall was actually assigned to a combined British-American engineer unit sent in to clear obstacles H-3 but that landed in the wrong place below St. Mihiel. The unit cleared the mines and obstacles, figuring some other lost unit would be doing the job they were to do (and did). The map story was a cover story for security.

tuesday — 6 june 1944

D-Day! The invasion of Festung Europa was on at last!

WEDNESDAY — 7 JUN 44 TO SUNDAY — 9 JUL 44

Two weeks after the invasion of Europe by the Allies, Lt Hall . . . [returned and] relieved Capt Gonzalez at Langar. The Captain returned to Riseley where he remained with the Company except for an absence of several days.

Once the attack was secured and the Allies established on the Cherbourg [Cotentin] Peninsula [Cherbourg was the major port city the Allies wanted to seize early in the campaign] . . . the airdromes were closed as paratrooper stations. All equipment was returned to proper depots. There was no work for the SOS [Services of Supply] Detachments. Orders to return to Riseley were awaited, and while we waited we loafed gloriously. It even became boring after a while for some of us; we wanted to fight a war.

The waiting lasted until Thursday, 6 July. That day we were ordered back.

Officers, Company B, 1303rd, Port of Embarkation, England, July 1944: Lt. Earl E. Hall (front left), Lt. Weyman M. Moseley (front right), Lt. D. M. Geesey (back left), Capt. Manuel F. Gonzalez (back middle), Lt. C. J. Deveikis (back right). Courtesy Earl E. Hall

The next day was spent in preparation for the return and on the 8th the officers began bringing back their groups. Everyone returned having earned a "Well Done." The last to return was Lt Deveikis. He came back on Sunday morning, 9 July, and found the Company packing. The rumors had been right: as of the 17th of July the Regiment was attached to Third Army, secretly forming under command of Lt Gen PATTON. That was a thrill. The complications of our second POM were gone into with a will. On Thursday, the 13th, we moved the Company into a nearby storage hangar, policed our camp site, completed packing, and waited for orders to report to the Staging Area for shipment to Europe, to France.

STAGING AREA — HILTINGSBURY, HANTS [NEAR SOUTHAMPTON], VU8742, ENG GRID. ARRIVED 1800 17 JUL 44 — DEPARTED 1230 23 JUL 44.

MONDAY — 17 JUL 44 TO SUNDAY — 23 JUL 44

Company "B," along with Headquarters 1st Battalion and Companies "A" and "C," departed Riseley, Bedfordshire, 1–1/2 mi N, wL4984, Eng Grid — for three months its English home station variously known as AAF Station 541 and Melchbourne Park—at 1020 hrs, enroute to the Southampton Port of Embarkation by rail and arrived at Eastleigh, Hamptonshire, at 1620 hrs. From there the trip to Hiltingsbury, vU8742, Eng Grid—Camp C-6 of the Marshalling Area—was made by truck ending at 1800 hours at destination.

Here a new experience awaited us: our housekeeping was to be done for us by the static staff (British combat veterans) while we only completed POM requirements for the short sea voyage which, in our case, consisted of exchanging the men's '03 rifles for M-1s. Beyond that we only ate, slept, rested, and tried to keep clean for the next six days.

Overstrength by three enlisted men, we dropped them on the 18th leaving the Company at Table of Organization [T/O][15] strength: 5 Officers and 168 enlisted men. Present were four officers and 137 enlisted men absent were one officer and 31 enlisted men—two enlisted men on the Battalion Advance Party, the rest with Lt HALL on the motor vehicle convoy, traveling separately.[16]

ABOARD HMS "DUKE OF WELLINGTON" ON THE ENGLISH CHANNEL.

EMBARKED 1600 23 JUL 44—DEBARKED 1710 24 JUL 44.

SUNDAY—23 JUL 44

The Company received Movement Orders at 0800 hours and departed the Marshalling Area by truck at 1230 hrs, arriving at the Southampton docks at 1330. We boarded the transport, HMS "Duke of Wellington" (6,000 Tons),[17] between 1545–1600 hrs. With all the troops aboard the ship moved out to the mouth of the English Channel to await the convoy for France and anchored at 1830 hrs.

CHAPTER FOUR

Normandy and Brittany

JULY–AUGUST 1944

When the 1303rd landed in France on 24 July, Patton himself had only been ashore eighteen days. In another four days Bradley would decide to unleash Patton and the Third Army. Becoming operational on 1 August, the regiment was just in time to be involved in the German attack at Mortain — Operation Lüttich. As the 1303rd came ashore and moved inland to its first bivouac site, passing Ste. Marie Eglise and the detritus of war, the engineers could hear the rumble of artillery and the crack of lighter weapons.[1]

Moving inland, they first headed northwest up the Cotentin peninsula toward Cherbourg before swinging south into Coutance. From there ran only one major road, which passed through the city of Avranches. As the Americans moved toward Avranches, the German Seventh Armee prepared to implement Operation Lüttich. Planned by Generalfeldmarschall Guenther von Klüge, commanding OB West — all German forces in northern France — with personal modifications by Adolf Hitler, Lüttich involved an assault intended to gain control of the road. This would split the American forces in two, bottling up most of the Third Army, without the opportunity for resupply, to be dealt with in detail later.[2]

The Mortain offensive was late in starting, and on 7 August as the Thirteen Third moved into Avranches along the single road (which was probably twenty- to twenty-four-feet wide, the standard width of the Routes Nationales),[3] the men had to deploy eastward as infantry. Fortunately, after a strong fight by elements of the VII Corps, the German advance was halted short of its goal. One of the key units defending what, by all accounts, was the critical point — Hill 317 outside Mortain — was the 120th Infantry Regiment, 35th Infantry Division (Kansas National Guard). The road remained open to the Third Army as it moved into Brittany.[4]

As the regiment came ashore in Normandy, it entered the "theater of operations." Generally, a theater of operations "comprises land and sea areas invaded . . . , including what is necessary for administrative establishments and agencies pertaining to the forces in the theater." It is divided into at least two parts, the "combat zone," in which combat operations are in process, and the "communications zone," in which support forces operate. The rear-most area of the communications zone is the "base section." Britain was the base section of the communications zone for operations

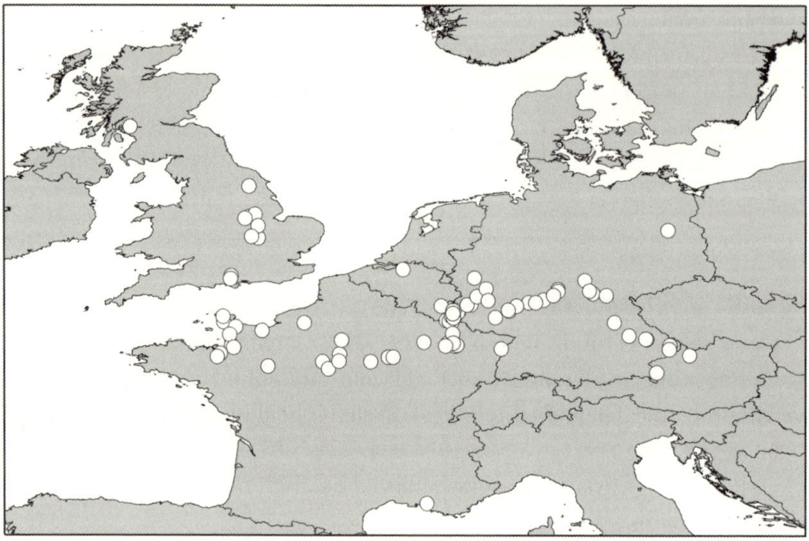

Europe. Courtesy Paul A. Lorah

on the Continent. At first, the area of the landing beaches in Normandy was the "advance section," but as the armies moved deeper into France, the advance section moved as well, and the beaches became an "intermediate section." Engineer general-service regiments were supposed to function in the communications zone, not the combat zone. The difference between the various operational regions in which the 1303rd served illustrates the unique nature of its wartime record.[5]

The combat zone itself is divided. It is easiest to start at the "front line," where enemy troops are engaged. This is apportioned into several "division areas of responsibility" (for example, the area assigned to the 4th Armored Division or the 35th Infantry Division), behind which lies the "corps area of responsibility" (here would be the XII Corps area or that of the XX Corps). Behind this, extending to the border of the combat and communications zones, was the "army area of responsibility." The Third Army was the westernmost and southernmost Allied command landed at Normandy.

NORMANDY, FRANCE

TRANSIT AREA — NORMANDY, FRANCE.

ARRIVED 1945 24 JUL 44 — DEPARTED 1200 25 JUL 44.

MONDAY — 24 JUL 44

Our ship sailed in convoy at 0330 hrs. . . . It was a very slow trip as constant vigil for mines was necessary. The water was plate-smooth, the weather fine.

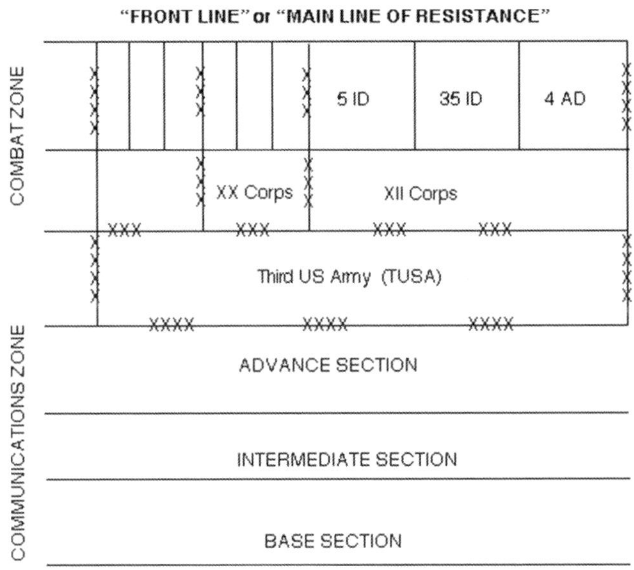

Organization of the Theater of Operations. Courtesy JCF

At 0900 hrs, some 17 hours after boarding the ship, the coast of France hove into view. The outline of Cherbourg was clearly visible with glasses. . . . The scores of vessels were protected by barrage balloons as they awaited their turn to unload. The company went ashore on Omaha Beach at 1700 hours and . . . proceeded to the transit area to spend the night.[6]

What a procession that was! With the Colonel at the head of the column of mobs that staggered off at a pace that was a cross between "quick" and "double" time, we took off. Everyone was in full field equipment. There were no halts. There was no formation. There were no commands. The distance was about six miles. The column must have stretched the whole distance from beach to bivouac area. The head of the column staggered in around 1945 hours; the tail dragged in about one hour later. "Column of bunches, drag arms and follow me" might well have been the command for that march.

That night we got our first real taste of ack-ack [antiaircraft artillery, or AAA] fire when the gun near which we slept opened up on some German planes. It was joined by the other guns of the battery and made a respectable noise even though nothing was hit. In the distance could be heard the booming of the front-line guns: we were less than ten miles from the fighting, a little north of Carentan.

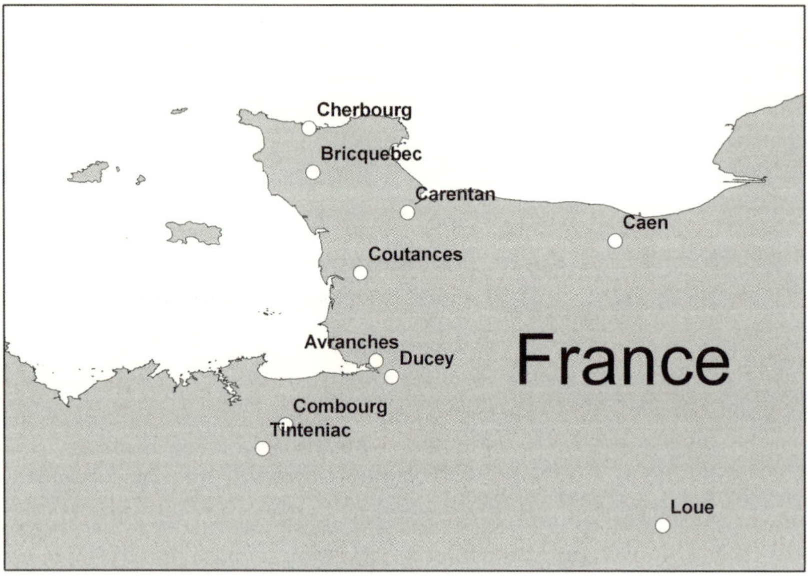

Normandy and the Cotentin Peninsula. Courtesy Paul A. Lorah

BIVOUAC 1—QUETTETOT, 1/2 MILE WEST 098067 FR LAMBERT 1. THE MAP SERIES IN USE IN THIS REGION WAS BASED ON THE FRENCH LAMBERT 1. THE MAP ON P. 29 IS AN EXAMPLE OF THIS. COMPANY B IN FACT MAY HAVE USED A COPY OF SHEET 54 (CHERBOURG-ROUEN), EUROPE ROAD MAP.

ARRIVED 1400 25 JUL 44—DEPARTED 2345 1 AUG 44.

TUESDAY—25 JUL 44

After a breakfast of K rations following a supper of K rations the night before and before a dinner of K rations after-wards,[7] and a four-hour delay, we departed the transit area at 1200 hours and proceeded by motor convoy (a Quartermaster Corps truck company) to our first bivouac in France, close to Bricquebec, and to our introduction to the apple orchard country and the hedgerows of Normandy.

Enroute we drove through the gutted towns of Montebourg and Valognes. It was the first sight of war for most of us. We caught a brief glimpse of the pile of rubble that was a town and an occasional stark fragment of a wall. The people looked poor, tired, and dirty, yet the men said they were friendlier then the British had ever been. They passed out a lot of "cidre" to truckloads of thirsty soldiers while we waited for a guide. The first experience with the French was not unpleasant.

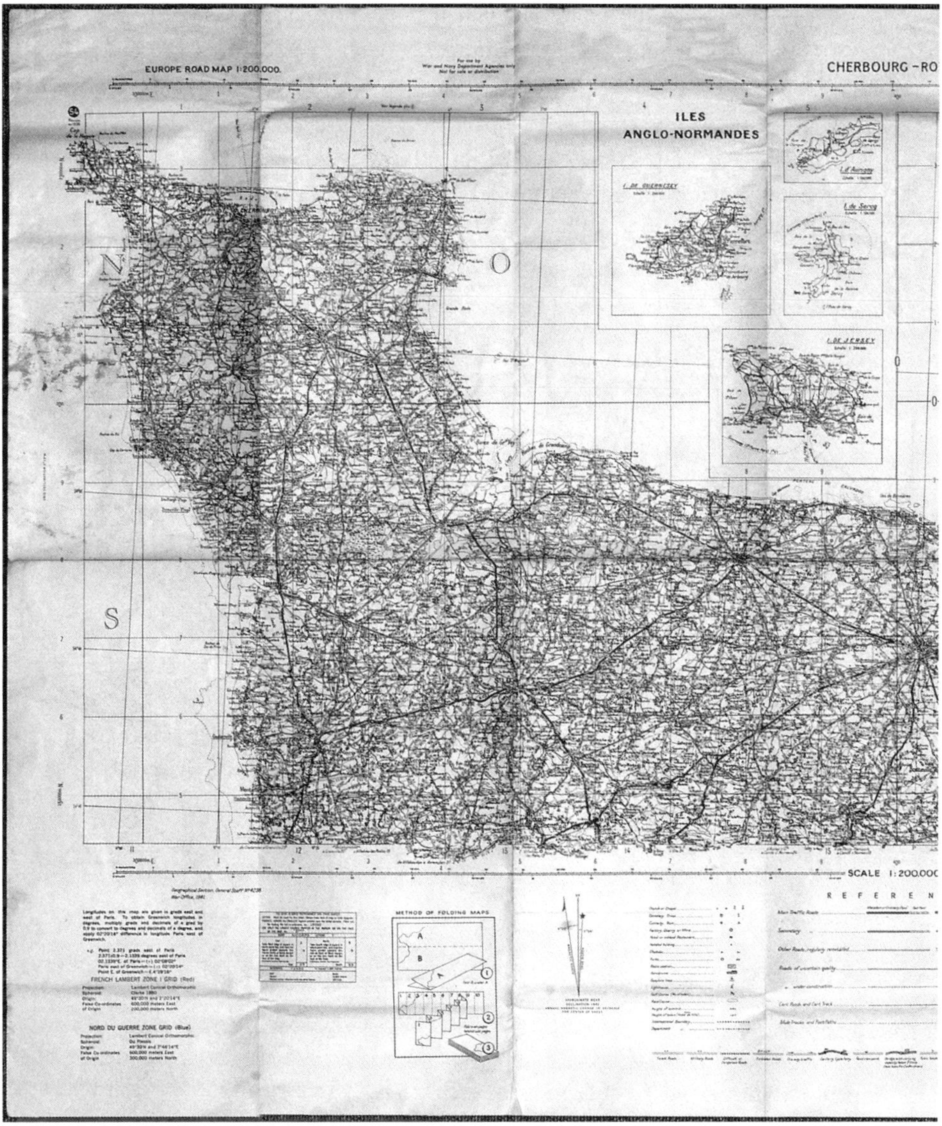

Left Half of Sheet 54 (Cherbourg-Rouen), Europe Road Map, Using the Lambert and Nord du Guerre Schemes, from Army Map Service, 1942. Courtesy Doug Bekke

WEDNESDAY — 26 JUL 44

The advance party . . . and the motor vehicle convoy arrived . . . with . . . our equipment provided, as bedding rolls and duffle bags, tools, and especially "Dunbar's Diner"—the kitchen, which promised hot cakes for the following morning.[8]

. . . .

FRIDAY — 28 JUL 44

The Battalion was assigned a reconnaissance mission.[9] The Company sector is bounded by Road 206 on the north, 200 on the south, 325 to the east, and the coast on the west. . . .

SUNDAY — 30 JUL 44

The 3d Platoon began work on the back road from Regimental Headquarters to Quettetot using six borrowed dump trucks.[10] The bathing in Cherbourg [begun on 28 July at a quartermaster shower unit] continued until yesterday when the MPs [Military Police] ran out a group for being in town without passes and armed. . . . [They enjoyed] the coffee and doughnuts [at] the "Club Victoire," the first . . . recently opened American Red Cross Service Club on the Continent.

. . . .

MONDAY — 31 JUL 44

Still sweating out our first Third Army mission.

THIS "FIRST MISSION" INVOLVED keeping open the Third Army's lines of communication and supply. The regiment's men patrolled roads to find and repair problems to ensure two-way traffic, especially on bridges. During their service in five European campaigns (Normandy, northern France, Rhineland, Ardennes, and central Europe), the 1303rd erected hospitals and prisoner-of-war enclosures, prepared cemeteries, and assisted in several river-assault crossings, often under artillery fire, as well as building bridges and maintaining roads.[11]

TUESDAY — 1 AUG 44

Ordered late in the day to move out on our first task, we departed at 2345 to travel all night in unbelievably heavy traffic.

BIVOUAC 2—COUTANCES, 4 MILES SOUTH 2649, FR LAMBERT 1.

ARRIVED 0700 2 AUG 44—DEPARTED 1000 7 AUG 44.

AT THE END OF July 1944, the VIII Corps (under Patton's control since 28 July) began to shift emphasis from "positional warfare in Cotentin to wide open exploitation in Brittany" as Patton began implementing his style of combat. Maj. Gen. Troy Middleton (commanding general of VIII Corps) found this unfamiliar and quite upsetting, as did the air units of the XIX Tactical Command that had been supporting VIII Corps operations. Like the Third Army, XIX TAC became operational on 1 August, under the command of Brig. Gen. Otto P. Weyland. As Patton's troops began campaigning with a cavalry mindset and dash in place of a slower, more methodical infantry style, XIX TAC air crews were asked not to strike bridges, which helped the Third Army advance faster. Patton saw no reason to have to stop to fix what the air force had blown up if there was no need to destroy it in the first place. Once he got the Germans to fall back, they did not stop long enough for air crews to mount bombing strikes and usually did not wait long enough to blow the bridges themselves.[12]

To enter its operational area in Brittany, the Third Army moved from Coutance down a single road through the small town of Avranches, between the See and the Selune rivers. At the Selune the road divided into three routes, one west into Brittany, one south toward Fougères, and the third east to St. Hilaire du-Harcouer. As the 1303rd passed through Avranches, the German lines were approximately six kilometers from the ocean, about two kilometers from the road. On 7 August, as Patton expected thanks to ULTRA intercepts, the Germans launched a substantial attack aimed at cutting through to the sea. ULTRA revealed that the 116th Panzer Division was ordered to "counter thrust towards Avranches" and that Luftwaffe Air Tasking Orders and artillery were focused on cutting the Third Army's bridges and lines of communication. The attack came about the same time the 1303rd was moving through Avranches. The regiment was halted to deploy as infantry in case the Germans managed to break through American lines. Fortunately, the assault failed.[13]

WEDNESDAY — 2 AUG 44

The bivouac area we occupied, left by the retreating Germans two days previously, was littered with abandoned equipment, explosives and ammunition, and clothing.[14]

Assigned road reconnaissance, maintenance, and repair of N800 from road junction [RJ] N800-Gc2 (South [of] Lessay) to Ferrieres; N171 through Coutances to RJ N171-Gc7. Through Coutances priority I; remainder, priority II.

Lt Moseley and his 3d Platoon were assigned the task of building a flight strip for "Eagle Main" for Lt Gen Bradley's use vicinity St Sauveur. Departed on Temporary Duty today.

THE 1303RD EGSR WAS assigned to and controlled by Third Army headquarters. Lacking telephone connections, 2nd Lt. Robert McRae, the regiment's assistant engineer officer, became the liaison officer to the army's engineer section. Initially attached to the VIII Corps, either the regimental commander or executive officer went to army headquarters daily. While the 1117th Engineer Combat Group (ECG) was responsible for the VIII Corps Area, the 1303rd was responsible for assisting the 1117th ECG as well as for "keep[ing the] lines of communication open in the entire army area in support of VIII Corps." Later the executive officer frequently visited the engineers of the VIII Corps and the 1117th. When the regiment was attached to the XII Corps, another officer served as liaison with the corps engineer and later with the 1137th ECG when it replaced the 1117th in support of the XII Corps. After operations became "comparatively static the liaison officer to XII Corps was reassigned to liaison with Third Army, and the 1137th Group was contacted daily. The 632nd Engineer Light Equipment (LE) company, commanded by Capt Paul D Orleman was attached under Letter orders, HQ 3 USA, file AG 322, Troops Assignment No. A-13 of 1 Aug 1944."[15]

THURSDAY — 3 AUG 44

Clearing Coutances of rubble from wreckage consists of demolishing dangerous overhanging buildings and bulldozing the debris out of the streets. There are some craters to be filled.

Flight strip — Standing timber in the approach zones removed, hedges in the area of the strip cleared, and terraces in the apple orchard leveled.

FRIDAY — 4 AUG 44

Our area assignment was changed: the new area centers about Gc7 from Coutances to Brehal and Gavray.

Flight strip — One-quarter cleared and graded.

Road maintenance — Clearing Route Gc7 Coutances, Lengronne, Brehal and the road Lessay, Ferriers, Coutances. Coutances is nearly cleared as of 1800 hrs, and will be completely cleared by 2100 hours today. Two buildings and one road block were demolished today.

Bridge maintenance — The 1st Platoon detail maintaining the Bailey Bridge South [of] Road Junction Gc72-Gc57 was bombed and strafed during the night without incurring any casualties.

At the Command Post, we "rescued" the uninjured pilot of a P-51 after he bailed out following an engine failure. The plane was burned.

The German dead along the roads and in wrecked and burned out vehicles in our area are a thick and sobering sight.

SATURDAY — 5 AUG 44

Seven dump trucks were attached from the 764th Engineer Dump Truck Company, a colored unit.[16]

The 1st Platoon continues to improve the supply route Brehal-Coutances-Lengronne. On Gc2 the 2d Platoon cleared rubble, filled a trench, patched a crater surface, and removed a block.

Bridge maintenance — A new approach is being cut to expedite the north end of the Bailey Bridge at Road Junction Gc20-Gc49. The 1st Platoon was relieved of maintaining the Bailey Bridge at Gc72-Gc57.

Flight strip — With the removal of projecting roots and ant hills the grading of the flight strip is 65% complete.

The Company was relieved of road maintenance responsibility by elements of the 341st Engineer GS Regiment.

SUNDAY — 6 AUG 44

Flight strip — At 1500 hours the flight strip was completed and immediately put into operation. Maintenance responsibility was turned over to the relieving unit, Company B of the 341st Engineers.

Road maintenance — no responsibility.

Bridge maintenance — no responsibility.

The 764th trucks were detached from the regiment and sent back to Quettetot to bring forward the 1301st and 1306th Engineer GS Regiments. Shades of Camp Ellis!

BRITANNY, FRANCE

BIVOUAC 3 — DUCEY, 1 MILE NORTH T3210, FR LAMBERT 1.

ARRIVED 1500 7 AUG 44 — DEPARTED 1000 10 AUG 44.

MONDAY — 7 AUG 44

Opened new CP at 1500, after a five hour ride from the old area, where we received a new assignment: Gc5 through Avranches, N799 from La Chapelle to St Hilaire.

While hunting Pfc Gryder's truck lost from the convoy, Lt Deveikis with Pfcs Bates and Lanzisera drew sniper fire just outside of Granville. In Avranches traffic was slowed to a bare crawl by enemy aircraft overhead flying a bombing mission. Intense AAA fire downed one plane which they saw falling in flames. About midnight bombs fell so close to our bivouac area that we felt the concussion. Parts of Avranches badly bombed.[17]

We entered Brittany with this move. The war does not seem to have touched this dreaming, contented countryside: the shops are open and full of stock; the crops are in good condition; the people seem well fed. In St Hilaire, though, where our artillery is located today, (a short distance from our bivouac) embers of bombed houses were still smoldering and smoking this evening.

TUESDAY — 8 AUG 44

Enemy aerial activity, while not great, nonetheless caused much speculation as to the extent of damages to our road network during the night. No damage found in the morning. Roads are being maintained, craters filled. Wrecked vehicles removed from N799. Craters in N807 filled and surfaced with shale. Cleared rubble and burning timbers from the streets of

St Hilaire. Rock crusher put in operation. Roads Gc78 and 103 cleared of wrecked and burned vehicles. Reconnaissance of and recommendation for removal and replacement of the 110' Bailey Bridge at St Hilaire made.

We are close enough to visit Mont St Michel at twilight before the tide came in.[18]

WEDNESDAY — 9 AUG 44

The general situation is interesting. We relieved "C" Company on road work so they could build 300' of floating Bailey near Pontaubault. "A" Company is guarding the dams at Barrage de Vexins on the Selune River. The dams, if broken, would flood the entire country and wash out all bridges to Avranches. Having relieved "A" Company of their road responsibility, too, we are thus maintaining the entire Battalion sector. All roads have been completely cleared except for four Mark IV tanks which we must have a D-7 [a caterpillar tractor] to move.

The seven 764th dump trucks returned to the Company today.

Third Army's rear boundary was moved back because of heavy bombing attacks. Army's advance, however, is not slowed down.

BIVOUAC 4—COMBOURG, EXO397, FR LAMBERT 1.

ARRIVED 1200 10 AUG 44—DEPARTED 1000 13 AUG 44.

THURSDAY — 10 AUG 44

Departed old area 1000 hrs. New Command Post located 2 Kilometers east Epinac. The night of 9–10 Aug uneventful. Air activity light, but no damage to roads. Early morning hasty reconnaissance revealed no obstacles in road net. One Bailey Bridge, D/S[19] 60'1, cl 40, in Dol on N176 spanning a single blown railroad span. The span was shortened by the 3d Platoon to 40'L and the bridge reclassified to 70.

FRIDAY — 11 AUG 44

Removed logs from shoulders of N796. Filled holes on shoulders of N796 and N155. Surface repairs made on filled craters N795 and Gc83. Removed German guns on Gc83 and wrecked vehicles on Gc85 and Gc4. Blasting heavy concrete obstacles in Dol and maintaining the Bailey Bridge there.

The first Battalion had attached for duty one platoon of the 632d Engineer Light Equipment Company.

Tec 5 Roberson drove Lt Deveikis and 1st Sgt Douds to visit Mont St Michel.

SATURDAY — 12 AUG 44

Removed tank obstacles — one north, one east, and one west — out of Dol. Repaired treadway and filled rut on approaches to Timber Trestle (TT) bridge, in Dol. Removed tank on N155. Filled fox holes on shoulders of N155. Demolished dangerous walls in Avranches. The Bailey Bridge in Dol and floating Bailey Bridge in Pontaubault being maintained, as are all the roads in the Company area.[20] Relieved of responsibility for Avranches and vicinity at 1400 hrs, the 1st Platoon returned to the Company.

The past two days have been made for aerial war. Formations of bombers and fighters, the largest since D-Day, have been pounding German communications. Some 600 locomotives and 5,000 railroad cars have been destroyed.[21]

BIVOUAC 5—TINTENIAC, 1/2 MILE SOUTH X0461, FR LAMBERT 1.

ARRIVED 1300 13 AUG 44—DEPARTED 1000 16 AUG 44.

SUNDAY — 13 AUG 44

The Company met and fell in love with Mmle Jacqueline of Beton when we bivouacked in her father's fields.[22]

This is the third day of tremendous air assaults on the enemy. In the new area our reconnaissance continues. The impetuous Battalion Commander sent a message to the Company to stand by for another move in the morning. "We head for Paris," he wrote.

MONDAY — 14 AUG 44

The Company performed no Engineer Operations today. Reconnoitering, Lt Moseley found a German munitions dump containing in addition to other explosives some 40 Ton[s] of German TNT. To date all of our demolitions have been done using German explosives, captured dynamite and shaped charges. Results have been excellent, especially with the shaped charges with which directional demolition is possible.

The Sergeant of the Guard in Company "C" shot and killed a man of that company when he refused to give up his loaded rifle with which he was, while drunk, threatening the men around him. An investigation has been ordered to close the incident formally.

DURING AUGUST, OPERATIONS INTENDED to keep open the lines of communication were often conducted under enemy air attack and small-arms and machine-gun fire. Given the paucity of transport vehicles, each company had to be able to move itself, but the 764th Engineer Dump Truck Company remained attached to the 1303rd, and its vehicles gave adequate lift to the line companies. The T/E assigned one prime mover and two 8-T trailers, but these were inadequate to move the four R-4 tractors assigned. When two D-8 tractors were attached without transport, "the ability of this unit to move and to fully utilize its available equipment on construction work" was degraded. The equipment from the attached 632nd Engineer LE Company was essential to move the D-8s. "The two 1/2 yard machines [cranes] assigned by T/E and two 1/2 yard machines [cranes] in addition to 1/2 yard Quickway Crane available from platoon of Light Equipment Company are not always sufficient to satisfy the required [*sic*] of six companies when they are all busy at the same time. There is also often a shortage of welding and cutting equipment to dismantle wrecked bridges and to cut and weld steel for new bridge construction. . . ." Not surprisingly, "the vehicle T/E of a General Service Regiment not being adequate to move the entire Regiment," when the 1303rd changed bivouac, it usually left one or two companies and equipment to follow later.[23]

TUESDAY — 15 AUG 44

Removed wrecks from N177, southwest Gougeses; removed rubble from the streets in Rennes on the approaches to the bridges built by the 2d Battalion of this Regiment. The Company was warned to have a full load of gasoline for a long move tomorrow.

THE START OF AIRBORNE landings in southern France heralded the campaign of the Sixth Army in Operation Dragoon. The Allied high command finally announced the formation of the Twelfth Army Group under Lt. Gen. Omar Bradley, composed of Patton's Third Army, Lt. Gen. Hodge's First Army, and the French 2nd Armored Division. The deception campaign aimed at convincing the Germans that the main D-Day landings would take place at Calais had worked. Because Patton was commanding the fictitious First Army Group that was poised to make this landing, the announcement of his command of the Third Army was delayed and efforts made to keep the Germans thoroughly confused.

CHAPTER FIVE

The Rush across France

AUGUST–SEPTEMBER 1944

After getting away from the beaches and out of the hedgerows, Patton's forces moved into Brittany. Very quickly Bradley and Patton determined that they did not need the entire Third Army to clear the Germans from the peninsula and the major ports of Brest, Lorient, and St. Nazaire, the U-boat base. Wheeling south and then east, the Americans began their movement into the heart of France, with the newly operational XII Corps on the right. This opportunity came far earlier than anticipated, and Patton's speed of movement in response was unexpected by the Allied high command (Eisenhower, Montgomery, and Bradley). They saw that, as Patton surged eastward and Montgomery's XXI Army Group and Lt. Gen. Courtney Hodges's U.S. First Army also moved out of Normandy, they would trap the opposing German Seventh Army in a great sweeping envelopment, which became known as the "Falaise Pocket." Closing the "pocket" took too long, however, and many German soldiers escaped, though without their equipment.[1]

As the Third Army continued its advance, the XII Corps happened to have the 1303rd EGSR on *its* right flank, making this engineer *general service* regiment the flank guard of the entire army. To its south were Germans, mostly below the Loire River. The U.S. Seventh Army, operating in the south of France, was quite far away. While German units in western France began their own eastward movement to avoid being cut off by the converging Allied armies, they remained potent fighting forces.

BIVOUAC 6—LOUÉ, Z323327, FR LAMBERT 1.

ARRIVED 1600 16 AUG 44—DEPARTED 0800 22 AUG 44.

WEDNESDAY—16 AUG 44

Departed the old area 1000 hours and proceeded 100 miles to St Georges du Bois. This is the longest move to date. There was much German equipment wrecked and headed east, indicating they had been caught retreating. Laval and LeMans are wide open towns: the shops and streets were packed with people and appeared oblivious to the war. We have come 200 miles since Quettetot (15 miles from Cherbourg) in the 22 days since 25 Jul.

The 1st and 3d Platoons and Company headquarters are in the new

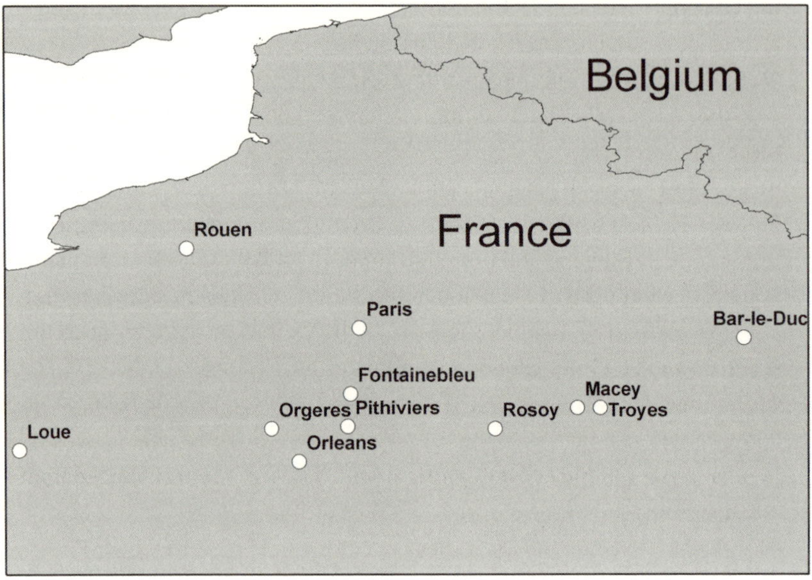

Company B Bridge Locations in Central France. Courtesy Paul A. Lorah

area. The 2d Platoon remained with the detachment of heavy equipment operators near Betton. Trucks were dispatched at 1800 hours to bring them forward.

Our southern flank is entirely unprotected.[2] The Regiment is together in this area. Our work depends upon the situation: i.e., we must find out where our flank ends and the German flank begins. Generally the enemy is supposed to have been driven out to below the Loire River but small pockets remain to be cleaned out of our area. Patrols and interior guard placed on the double alert.

THE NEW COMMANDER OF XII Corps, Maj. Gen. Manton Eddy (replacing Maj. Gen. Gilbert Cook, who was medically unable to continue service), was quite concerned about the right flank and hard pressed to understand Patton's relaxed view of the situation. Brig. Gen. Paul D. Harkins, Patton's deputy chief of staff, told the story that Eddy called the army commander saying, "Listen, your G-2 [intelligence officer] tells me there are 90,000 Germans on my right and 80,000 on my left and I've only got four divisions." Patton replied, "Ignore the ba____ards, go ahead." He later called General Weyland and told him to watch the army's flanks from the air, and if any Germans were coming up the roads to let him know. "I was convinced our Air Service could locate any groups of enemy large enough to be a serious threat," Patton later declared, "and then I could always pull something out of the hat to drive them back while the Air Force in the meantime delayed their further advance."[3] Since he had access to ULTRA and Eddy (and the 1303rd) did not, Patton was able to be confident about matters. Eddy, the XII Corps, and Company B took a somewhat less optimistic view of the situation and were quite relieved that nothing happened. The result, as Eddy's command followed orders, was an advance that Patton called "the fastest and biggest pursuit in history." The XII Corps took the Germans by such surprise in Sens that some officers were still strolling the streets in dress uniform as American tankers moved into town.[4]

THURSDAY — 17 AUG 44

The 2d Platoon and all equipment arrived 1500 hrs.

We are guarding the bridge over the Sarthe River at Spay, and reconnoitering our area carefully. Lt Hall's 1st Platoon ran into scattered small arms fire without incident. Company D lost one man to a sniper while he was trying to turn his truck around after encountering German obstacle. Several other reconnaissance parties were halted a number of times by the Germans who have some patrols out.

The German scare is a very real one.[5] A French civilian reported 12 Germans nearby and investigating it on orders from the Battalion Commander we found none. Earlier, at 2200 hrs, a native reported to the Command Post that a two-way radio set was in the home of a "bad" woman in town. We are not concerned with this allied unity which was probably caused by too much wine on the part of the Frenchman. However, we are concerned with the gendarmes' report that four German machine guns are located south of Clere, two of them in a cemetery fronting the road. In Chateau du Loir they captured ten enemy soldiers. German officers are reported in Langlais. There is no American traffic there.

FRIDAY — 18 AUG 44

We continue to guard the Spay bridge. The guard observed a flashing light off the north end in a farm house. It flashed 20 or 30 minutes in varying intensity apparently according to a pattern. The flashing began [at] 2230 hrs.

Lt Hall took three of his men and a 31-man security guard to remove booby-traps from a captured dump containing 25 drums of oil at road junction N826-N155, northwest Orleans. They returned at 2130, after some six hours of hunting, having found neither dump nor booby traps in the vicinity. While checking the site Lt Deveikis made a hasty reconnaissance of Orleans and found a most attractive city somewhat damaged by the war, all of its bridges blown. The freshly captured city was garrisoned with one US tank and its four tankers.

SATURDAY — 19 AUG 44

Reconnaissance showed the road net between La Fontaine and St Calais on Gc20 and Gc2 suitable for one-way traffic only: there are sharp curves and narrow streets through the villages enroute.

Continue bridge guard on Sarthe River at Spay and Ferce, which was added today. The flashing light was again observed last night by the Spay guards, coming from the farm house 600 yards north of the bridge. It is the first house on the right as one leaves the bridge. French civilians in the vicinity were questioned. They are noncommittal, say it is the FFI [Forces Francaise l'Interieur — Free French Forces, the Resistance]. Two children say occupants of the house are "no good."

. . . .

MONDAY — 21 AUG 44

We were relieved of bridge guard at 2400 hrs, midnight.

BIVOUAC 7 — PITHIVIERS, W 8970, NORD DU GUERRE ZONE. [NORD DU CUERRE ZONE REPRESENTS A NEW MAP SET.]

ARRIVED 2000 22 AUG 44 — DEPARTED 1000 24 AUG 44.

TUESDAY — 22 AUG 44

Departed the old Command Post vicinity of Mans 0800 hrs, proceeding to rendezvous point vicinity Cloyes, and arrived there at 1130. Ordered to build a Prisoner of War [PW] enclosure for Army in Nemours where, although the tactical situation was not clearly defined this morning at Army Headquarters, it was deemed safe enough to work since the need for the pen is urgent. It is to be the largest yet built on the Continent.

THIS PW ENCLOSURE WAS "so far forward that the site had not even been captured from the Germans before the task was assigned." Construction was delayed, but preparatory work cutting posts, building guard towers and gates, and so on was completed. The site was selected on 23 August and in use by the twenty-fifth, though the pens and water tower were not completed until the next day.[6]

Captain Gonzalez with Pfcs Bates and Lanzisera in the command car and Capt Bolzer (1st Battalion Adjutant) with Pfc Reed in the jeep preceded the Company by one hour. They were to meet a representative from Army Engineer at the *Maire's* office to get the specifications. Lieutenant Hall departed thirty minutes after them with his weapons carrier as a point for the convoy which, under command of Lt Deveikis, departed Cloyes at 1300 hrs.

Not a great deal of military traffic was on the roads. In Pithiviers there were no signs of troops, but the streets were filled with cheering, goggle-eyed French natives shouting their welcome, "Vive l'Amerique." After checking the maze of routes through this good looking, fairly large town, the Company proceeded towards the last town before Nemours, Puisseaux. Enroute one of the 764th trucks swerved to avoid hitting a child, collided with another 6x6 truck, and had to be abandoned. There were no injuries.[7]

Some ten miles from Puisseaux a Capt Merzweiler halted the convoy. He was the officer supposed to meet the Company Comdr in Nemours. When the situation became known in Army Headquarters late in the morning, that Nemours was held by the Germans, Army could not contact us as we were on the march. Captain Merzweiler tried to halt us in Pithiviers but had missed the convoy. Everyone expected our forces to take the town by evening, so he cautioned Lt Deveikis to be doubly alert from there forward for scattered enemy and gave further instructions and the specifications for the pen to be built. It was agreed not to take the Company beyond Puisseaux if the point failed to meet us there as planned.

As the convoy started the point weapons carrier hove into view, flying low. Captain Gonzalez was aboard, as were his driver and assistant driver. Lieutenant Hall was signaling from the front seat "enemy in sight." The convoy halted. Captain Bolzer's jeep was directly behind the weapons carrier. Captain Gonzalez then told the events as follows.

Enroute to the Maire's office in Nemours, while about three miles south of the town, a German road block was detected. The vehicles stopped for a look and the Captains inquired of a farmer's wife by the roadside if the block was manned by the FFI. She assured them it was not but that Germans were defending it. The officers returned to their vehicles which they started to turn around when enemy automatic weapons opened fire on them.

Pinned down in a ditch next to the vehicles, they returned the fire with three M-1 rifles, two carbines, and a submachine gun for some 20 minutes. The enemy was using, it was estimated, several automatic weapons and about 15 rifles, and after about five minutes of frontal fire began to flank the cars to the left. At this Pfc Reed was ordered to turn his jeep around as a getaway vehicle, it being lower in silhouette than the command car and also because the faulty starter on the latter made the vehicle unreliable. Cautiously this was accomplished, the party boarded singly while returning the enemy's fire, and then departed the scene abandoning under fire the command car containing the company drafting set, some personal effects of the two enlisted men riding, and all of the clothing, equipment, and personal possessions of Capt Gonzalez and Lt Deveikis save what they were wearing. The vehicle number is W 20459617-S. Location: X235736, Nord du Guerre Zone. Casualties: none.

About five miles down the road the jeep stopped Lt Hall's weapons carrier, appraised him of the situation, and then both vehicles started back to halt the company. Enroute the retreating party met the advance reconnaissance elements of a tank destroyer battalion just entering the area. They refused to help recover the vehicle left behind because the flimsy Tank Destroyers were no match for a possible 88 mm gun behind the block. Shortly thereafter the Company was halted.

The convoy was turned around and returned beyond Pithiviers before bivouacking for the night. Meanwhile, Lt Deveikis with three enlisted men in his jeep went forward to Puisseaux where the FFI guarding the town warned him of the enemy ahead and refused him permission to proceed without armored help. The FFI reported the enemy present in strength. The only way to retrieve the abandoned car was to use a tank, and as yet there were no tanks in the area. He returned to the new CP at dusk.

IN THE STILTED LANGUAGE of the regimental S-3 report: "3 Kilometers W Nemours was an enemy road block. Capt Gonzalez came under small arms and machine gun fire, the command car was lost, escape was in a 1/4 ton truck. There were no injuries. The next day [23 August] the B Company weapons carrier came under fire from an enemy halftrack."[8]

WEDNESDAY — 23 AUG 44

Lieutenant Geesey with Capt Merzweiler located the PWE [prisoner of war enclosure] near Nemours during the day, the town having been entered and taken by armor during the night. We were still too far ahead of the combat elements, for it was not until 2400 hours that forward elements of the 80th Infantry Division (319th Infantry Regiment) moved into the vicinity of Pithiviers relieving the Company of responsibility for the area which we had held quite unwittingly with no more than a double interior guard. While seeking water the detail in a weapons carrier was fired upon by an enemy half track. Though the vehicle was hit several times, there were no casualties.

Whatever else there is, there is always Battalion Headquarters and Major [Fred G.] Mylius.[9] In the afternoon we received a message from him: he advised us to move back as our present location was "too hot."

Instead of heeding his kind advice, we cut 750 posts to length, hauled lumber for the towers from Chartres, nails and barbed wire from Engineer dump 3, LeMans. Hinges, carbide lamps, and water tanks were unavailable.

CONSTRUCTING THE PW ENCLOSURE was made more difficult by the shortage of concertina wire, all of which seemed to have "been sent to the Brest peninsula and single strand wire was all that was available." The design was modified to use "high wire fencing" instead.[10]

The news today is good. Paris is taken. The French people all around us were jubilant, and at night came to the bivouac area with wine

BIVOUAC 8—FONTAINBLEAU, X308742, NORD DU GUERRE ZONE.

ARRIVED 1300 24 AUG 44—DEPARTED 0930 28 AUG 44.

THURSDAY — 24 AUG 44

Opened new CP vicinity Nemours at 1300 hrs. A detail of carpenters was left at the old site to complete framing gates and towers for the pen.

FRIDAY — 25 AUG 44

Setting posts and stringing wire for the enclosure, which measures 1,000 × 1,000 feet, and clearing the site within the perimeter. Guard towers, gates, water tank tower, access roads, and 19 signs completed. 16 trash boxes are under construction. The PWE will be usable but not complete tonight.

SATURDAY — 26 AUG 44

The first Prisoners of War were received at 0800 hrs. The static staff arrived 1000 hours and started operation of this, the largest Prisoner of War enclosure on the Continent. The entire project was completed at 2200 hrs.

Personnel — Pvt Lange was hospitalized. Pvts Hughes, Williams, and Grover Whitaker went on Temporary Duty to Regimental Headquarters as guards.

SUNDAY — 27 AUG 44

Minor changes and additions were made to the pen at the request of its commander, a Lt Col Davis, FA [field artillery].

Personnel — Pvt Whitaker was taken sick, hospitalized.

BIVOUAC 9—ORGERES, W644578, NORD DU GUERRE ZONE.

ARRIVED 1400 28 AUG 44—DEPARTED 0930 31 AUG 44.

MONDAY — 28 AUG 44

Opened this new Command Post at 1400 hours vicinity Artenay.

Personnel — Pvt Puccio replaced Pvt G Whitaker on guard at Regimental Headquarters.

. . . .

WEDNESDAY — 30 AUG 44

Road maintenance continues.

Received movement orders 1200 hrs. Reconnaissance party in force departed CP 1330 hours to locate new CP vicinity Sense. The gasoline shortage is critical.

THE FUEL SHORTAGE RESULTED from General Eisenhower's agreeing with Field Marshal Montgomery's plan for the XXI Army Group to attack Antwerp and Arnhem in Belgium (Operation Market Garden). Since there was not enough gas for the push on Paris and for the American forces (Twelfth Army Group) and British-Allied forces (XXI Army Group) all to advance rapidly against the Germans, General Bradley's Twelfth Army Group was deprived of fuel and kept in place for some time. Patton chafed under this regime and did his level best to continue forward motion of some kind, if only an armored combat command. Before the gas gave out, the XII Corps and Third Army had crossed the Moselle River at Nancy, Metz, and Toul. One hundred miles from the Rhine, Patton could not move.[11]

Making matters worse, the army depended upon a supply-depot system for its logistics, which did not work well in support of Patton's mobile warfare. The depots were near the beachheads, hundreds of miles from the forward dumps that the troops depended on for their needs. Destruction of the French railroad network by bombing before the landings forced supplies to move by truck. The "Red Ball Express" started in August 1944, operating twenty-four hours a day, seven days a week, and ignoring the blackout orders. One hundred thirty quartermaster truck companies operated some 6,000 vehicles on dedicated one-way routes to the front. Leaving the heavy artillery west of the Seine, its trucks too were put to use moving supplies up from the depots, which remained near the beachheads in the communications zone throughout the advance. Allied forces used over 800,000 gallons of gasoline per day, of which the Red Ball Express used 300,000 gallons.[12]

By the end of August, ration reserves were depleted, little individual equipment was available as replacement, and medical and signal supplies were critically low. The XII Corps estimated it needed 200,000–300,000 gallons of fuel for a fifty-mile advance. On 24 August it received 31,000 gallons; the next day it had 75,000 gallons, much of it captured from the Germans at Sens. On 29 August, Combat Command A, 4th Armored Division, captured 100,000 gallons from the Germans at Chartres and was able to move forward to the Meuse River.[13]

BIVOUAC 10 — ROSOY, X7160, NORD DU GUERRE ZONE.

ARRIVED 1300 31 AUG 44 — DEPARTED 1300 3 SEP 44.

THURSDAY — 31 AUG 44

Opened new Command Post vicinity Sens at 1300 hrs, some four hours after departing the old area.

FRIDAY — 1 SEP 44

Road reconnaissance indicated little maintenance which is being performed. Infantry outposts are located on Gc56, Gc16, and N443 (St Maurice to Joigny). Chatillon not yet occupied by US troops. Cavalry reconnaissance Joigny SW on N455. There are no obstacles (tank blocks) on the road network.

. . . .

BIVOUAC 11—TROYES, Y2971, NORD DU GUERRE ZONE.

ARRIVED 1700 3 SEP 44—DEPARTED 1500 4 SEP 44.

SUNDAY — 3 SEP 44

Billeted Company in the St Juliene Dye Works Warehouse in Troyes at 1700 hrs, after a four hour trip.

Reconnoitered the two bridge sites and gathered some materials. Work will begin at 0730 tomorrow morning.

Headquarters 1st Battalion and 1st Battalion Medical Detachment attached for rations and quarters.[14]

BIVOUAC 12—MACEY, 1 MI SOUTH X1722, NORD DU GUERRE ZONE.

ARRIVED 1800 4 SEP 44—DEPARTED 1000 9 SEP 44.

MONDAY — 4 SEP 44

We moved to a new location because General Patton forbids his Army to live in buildings.

Bridging—Construction began at 0730 hrs.[15]

Site 1: Located in Troyes. Fixed bridge, steel beam, Cl 70, 65' long, one-way.[16] Removing old reinforced concrete beam structure and assembling materials. Job 15% done. Site 2: Located on N77 outskirts of Troyes. Fixed bridge, steel beam, Cl 70, 114' long, one-way. Placing cribs and assembling material. Job 15% done.

TUESDAY — 5 SEP 44

Bridging—Completion date for the two sites estimated as 2000 hours 8 Sep 44. When these bridges are opened there will be class 70 passage from east and north on Main Supply Routes [MSRs] N60, 77, and 19 through Troyes.[17]

Site 1: Old structure removed, site clear. Moved in five 36" I beams 75' long for crossing one span abutment to abutment. Work 35% done.

Site 2: Foundations and substructure 75% complete. Materials for superstructure and floor system assembled. Work 45% done.

Crane placing steel beams for Troyes bridge, France, 1944. Courtesy Dwight Gowdey

Setting steel and an air compressor on Troyes bridge, France, 1944. Courtesy Dwight Gowdey

Road maintenance — Cleared wrecked vehicles from N60, Troyes to Piney, N79, Piney-Vendeuvre.

WEDNESDAY — 6 SEP 44

Bridging — Site 1: Abutments completed. Two of four I beams hauled and cut to length. Job 50% complete.

Site 2: Foundations and substructure complete. Stringers in floor system being set as the hauling of 10" I beams is complete. Work is 65% done.

Road maintenance — Vehicles removed from streets of Troyes. Check reconnaissance of N19 to Vendeuvre completed and special report I submitted 3 Sep 44 substantiated as correct.

. . . .

FRIDAY — 8 SEP 44

Bridging — Site 1: Bridge completed 2130 hrs.

Site 2: Bridge completed 2130 hrs. We were one and one-half hours late in our construction, primarily because we ran short of decking on site 2 around 1700 hrs.

Installed curb and handrail on damaged arch bridge out-skirts Troyes.

Troyes bridge completed. This represents the typical bridge built by Company B. The cribs in the water support timber trestles, steel beams rest on the trestles, and the decking is attached to the beams. Courtesy Dwight Gowdey

The bridge is a safe cl 40, but cl 70 traffic has been using it all day. Previously reported crack in the first span has widened to a 4–10" fissure ten feet long. Recommend repair by bridging the first span using 36" I beams 65' long which are available at the railroad yard in the area. We cannot do this work because of movement orders for the morning.

Reconnaissance elements departed for Bar-le-Duc 1100 hrs, followed by the 2d Platoon at 1530 hrs.

CHAPTER SIX

Building Bridges

SEPTEMBER–DECEMBER 1944

The 1303rd EGSR had moved a long way from Ducey in about a month. Now the men would spend more time building bridges than moving. Many of the spans built by Company B would be permanent timber-trestle bridges that replaced a tactical Bailey Bridge. Whether other engineer units used the same method or not is unclear, but the bridge builders of B Company built their timber trestles underneath the Bailey, leaving it in use. Once their "TT" bridge was ready, they would stop traffic, remove the Bailey—a relatively fast process—and open their new bridge to traffic while putting on curbings, railings, and other finishing touches.[1]

The company's stay in Bar-le-Duc was surprisingly quiet. Once the unit moved into the vicinity of Toul, the men came under German artillery, mortar, and some small-arms fire. The engineers would build bridges and maintain roads under threat of enemy attack from this time until the end of the war in Europe. They also began regularly having to clear minefields and remove German demolitions, the one part of the 1303rd's training neglected at Camp Ellis. Even though several men were injured in the process, none were injured as they learned "on the job."

BIVOUAC 13—BAR-LE-DUC, 2 MI NORTH U087252, NORD DU GUERRE ZONE.

ARRIVED 1700 9 SEP 44—DEPARTED 1000 19 SEP 44.

SATURDAY—9 SEP 44

The new Command Post was opened at 1700 hours in the village of Fain, 3 Kilometers north Bar-le-Duc. No reconnaissance of the new area was made today.

The 2d Platoon had hauled 67% of the tactical bridging to Company "D" at Chalons where it is to be erected over the blown bridge before orders were received here to discontinue.

Battalion headquarters and Medical Detachment were detached from rations and quarters.

SUNDAY—10 SEP 44

Completed reconnaissance of roads in assigned area; report submitted. Our work is being done in eight places as follows. . . . Site 5: The 11'6" railroad underpass will be bypassed as shown on detail sketch when work at site 6 is

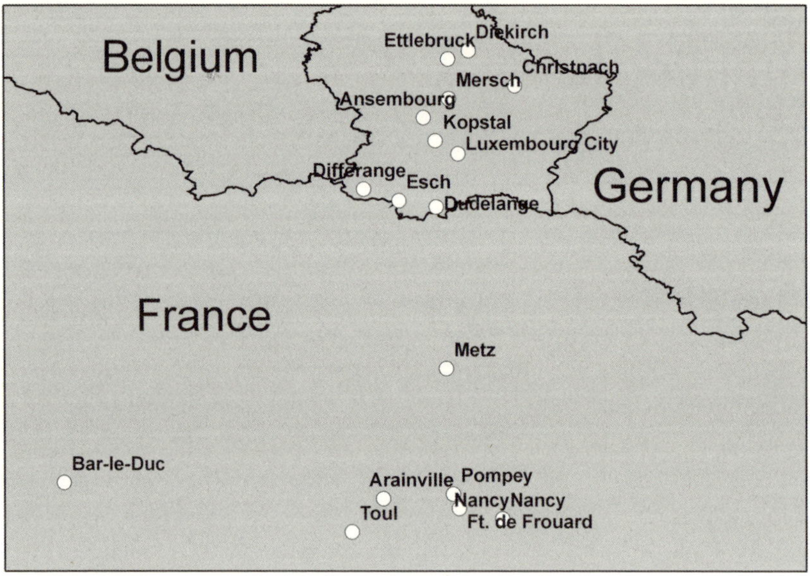

Company B Bridges in Eastern France. Courtesy Paul A. Lorah

completed. Site 6: The grade crossing is rough and difficult to negotiate at this site and its width is only 14', the restriction being railroad block signals. The crossing is being improved by building up and laying timbers between the rails and improving the grade and the approaches. Site 7: Work is in progress to strengthen bridge up to its classification of 40. As it was it could not be used for cl 40 loads. One additional bent under the I beam stringers butt-jointed between existing bents was needed. . . .

MONDAY — 11 SEP 44

Site 1: Work discontinued; 70 classification is not required by [a traffic] Circulation Diagram, Third Army Engineer, 8 Sep 44. Construction of footwalk continues. Site 2: Bailey Bridge parts stockpile unchanged. . . . Site 7:

Wo 165 w20 h22 [wooden bridge, length 65', width 20', height 22']
Cl 70 [class (load, one way) 70 tons][2]

Site 8: Work discontinued as cited Circulation Diagram does not require it.

TUESDAY — 12 SEP 44

Site 1: Footwalk 95% done. The width of the bridge was increased to two-way lane by adding an additional stringer under the deck along the edge of the bridge and building the walk on this addition.

. . . .

Site 4: Completed widening approaches and improving grade. Cinders dumped in and being spread to eliminate mud. Job 100% complete. Southbound bypass over Ic35 may be discontinued now; N66 approaching Bar-le-Duc is a two-way road.

Sites 5, 6, and 7 were completed 11 Sep. Site 8: 100% complete today. . . .

. . . .

THURSDAY — 14 SEP 44

Responsibility for Bailey Bridge parts at site 2 ended today. Completed improvement of bypass at Voie Sacree. Improving approaches on bridge South [of] Bar-le-Duc on N401 and the railroad crossing on a curve outskirts of town.

. . . .

FRIDAY — 15 SEP 44

Completed widening and improving the railroad crossing, including removal of all wrecked vehicles from shoulders.

Began asphalting approaches to bridge at Saudrupt.[3] Also, adding a second footwalk to the bridge at site 1 and two four-foot walks at site 2.

Reconnoitered roads. The railroad gate post is obstructing traffic on N77; removal contemplated, if approved.

Personnel — Pvt Netski assigned but not joined this date. Pvt Lange was transferred to Detachment of Patients and dropped from rolls.[4]

The 1st Battalion Company Commanders reconnoitered the area for bridge material for stockpiling it against an impending task. The dump will be established in our area.

SATURDAY — 16 SEP 44

Completed footwalks for both bridges. Completing approaches to Saudrupt bridge.

Maintaining bridge on N66 southeast [of] town. Cinder approaches being replenished; sheet piling driven at either end to retain fill.

The cinder approaches are adequate without asphalting.

Bridge timbers are being hauled to our bivouac area for stockpiling in a Battalion dump. Reconnaissance for materials [at] Toul and Nancy.

The railroad gate post carries four lines of two-way telephone wire. Its removal would interfere with signal communications and is not recommended.

Personnel — Pvt G Whitaker transferred to Detachment of Patients [hospitalized] and dropped from rolls.

. . . .

BIVOUAC 14—AVRAINVILLE, 1 MI SOUTH U6819, NORD DU GUERRE ZONE.

ARRIVED 1700 19 SEP 44—DEPARTED 1000 26 SEP 44.

TUESDAY — 19 SEP 44

New Command Post vicinity Toul opened 1700 hours after a seven-hour move. While reconnoitering for bivouac Capt Gonzalez and Lt Deveikis were in Pompey, freshly evacuated by the Germans. Artillery OPs [observation posts] in high heavily wooded hills near town are being used to direct the artillery dug in along the road (N407) between Marbache and Pompey. They were warned to beware of the German snipers remaining in the woods.

The bridge demolished by the Germans was a Ka L360' [360-foot-long stone-arch bridge] W20' H20', Cl 70. Four of the six spans are blown. These will be replaced with a two-way cl 40, one-way cl 70 Timber Trestle bridge on cribs founded on the rubble of the demolished structure. The nearby steel mill[5] was not investigated for possible beams. The enemy is in the next town across the river from the work site.

Personnel — Pvt Netski, our assigned but not joined replacement is hospitalized, effective 15 Sep.

WEDNESDAY — 20 SEP 44

The Pompey bridge was begun this date. Demolishing damaged portions of old bridge, assembling materials. The 1st Platoon is employed for security of site using emplaced machine guns.

Scattered enemy reported along N57 from Pompey south toward Nancy. About 1400 hours enemy light mortar fire on the bridge interrupted all work and personnel departed site after considerable activity and no small scare. Privates first class Roush and White and Pvts Freeman and Gainey were slightly wounded by shrapnel.

Earlier, about 1000 hrs, a German soldier surrendered to the outpost at Pompey. He was brought to the Command Post for interrogation. His account: four days ago his unit — 1120th Inf, 109th Division[6]—was holding the hill northwest Pompey when they were shelled out by our artillery. He and four others deserted. The others surrendered to the FFI, but he had a game leg and hid at the time. The PW stated he was Polish, conscripted for the *Wehrmacht* and had recently served on the Russian front.

Personnel — Pvts Jessie Bullock and Isham Wilson assigned and joined the Company.

THURSDAY — 21 SEP 44

Pompey bridge construction suspended due to enemy action. Small arms firing on the bridge at 1400 hrs. Patrols of the 28th Cavalry Reconnaissance Troop reconnoitering a crossing there. Enemy side of river shelled by US artillery. We are, however, hauling the timbers from Bar-le-Duc to vicinity Pompey.

Reconnoitered the 977' blown bridge at Pont-a-Mousson. Two companies of the 249th Engineer Combat Battalion are completing a bridge there using two Baileys. The east bank is heavily mined and some wire is erected. Nine head of cattle were killed last night in the field. Several Teller mines are visible on the river bottom at the crossing. Five or six American soldiers are lying unburied about 500 yds from the site.

FRIDAY — 22 SEP 44

Still cannot resume operations on Pompey bridge. Completed hauling up the timber dump from Bar-le-Duc. Hauling I beams from St Mihiel to bivouac area in lieu of site which is unapproachable with heavy trailers.

Personnel — Pfc Roush and Pvt Freeman were awarded the Purple Heart at the 106th Evacuation Hospital and returned for duty this date.

SATURDAY — 23 SEP 44

Work on Pompey bridge continues suspended. About 1830 hours the site was reconnoitered. 80th Infantry Division patrols on N57 reported small arms fire there (between Marbache and Pompey) today, but our reconnaissance party drew no fire. We continue to haul steel from St Mihiel. A shortage of timbers exists at the present time.

. . . .

MONDAY — 25 SEP 44

Resumed bridging operations at Pompey with one Platoon for security. Demolition crew blasting old piers. Moving timber and steel from bivouac to site. The river is rising.

BIVOUAC 15 — POMPEY, U8120, NORD DU GUERRE ZONE.

ARRIVED 1200 26 SEP 44 — DEPARTED 1500 27 SEP 44.[7]

TUESDAY — 26 SEP 44

The Company moved for the sake of convenience and closer proximity to its work site to the school house and auditorium in Pompey.

Preparing foundations for timber bents and fabricating the bents, also preparing approach. Estimate completion date 2000 6 Oct 44. Continue hauling timbers and steel from St Mihiel.

Reconnoitered the canal bridge which is completely destroyed including the abutments. The span can be bridged with 80' I beams. The bridge must be replaced to open N57 into Nancy.

Pompey bridge, France, 1944. Courtesy Dwight Gowdey

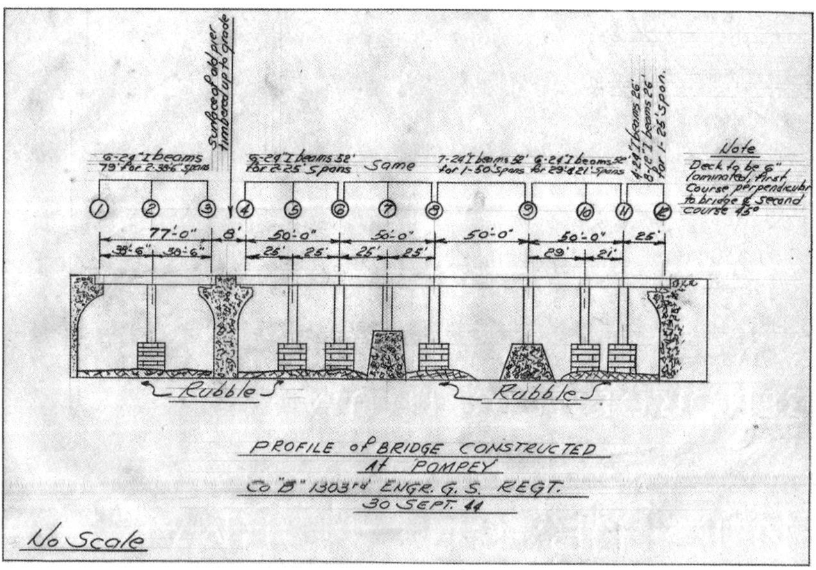

Drawing of the bridge at Pompey. Courtesy Earl E. Hall

BIVOUAC 16—FORT DE FROUARD, U8216, NORD DU GUERRE ZONE.

ARRIVED 1900 27 SEP 44—DEPARTED 1000 9 OCT 44.

WEDNESDAY—27 SEP 44

At the Colonel's insistence that it was too dangerous for us to live in Pompey, across the river from the body of our own troops, with only one bridge over which to cross in the event of a German counterattack, we moved across the river, to the vicinity of Champigneulles; more specifically, to the abandoned barracks of Fort de Frouard in the Maginot Line near Nancy.

The Pompey bridge is 10% complete. Five sites prepared for piers and two bents fabricated. Completed hauling steel and timber from St Mihiel. Bulldozing rubble to level bed for approaches and pier. Tractor unearthed 500 lb "unexploded US bomb in the debris of smashed buildings on near side of site."

. . . .

FRIDAY—29 SEP 44

All subfoundations at pier sites complete as well as four of the seven cribs, six of eight bents, and two of eight bents placed. Pacing 80' steel for two 40' spans at north abutment. Hauling timber and steel from north end to south end of the bridge and moving equipment the same way. Construction is going on from both ends of the bridge and the job is 25% done. Although at the start we had enough decking, such a great amount was used for cribbing that we now need 14,000 lineal feet 3 × 3" decking to complete the bridge.

The river is up 4.5' since the start of the bridge. Personnel—Pvt Golden assigned not joined the Company.

SATURDAY—30 SEP 44

All bents fabricated and eight of the 12 foundations ready for them. Three bents and sills placed. Six 80' I beams in place at north end. Hauling of material and equipment to the South end of the bridge is complete. Hauling additional decking to both ends of the bridge. Job 30% done.

Personnel—Pvt Kolbusz was transferred to Detachment of Patients and dropped from rolls. [He was hospitalized 14 September 1944 "for an appendicitis operation."]

SUNDAY—1 OCT 44

Foundations 1–7, 9, and 10 complete, including the erection of trestles and placement of sills. Subfoundations 8, 11, and 12 complete and bents for them ready. Steel for spans 1–3 in place with nailing strips. Bridging between the beams is done. Steel loaded for spans 4–6. Job 40% done.

Personnel — Tec 5 Nance fell from the bridge injuring himself seriously. He was hospitalized.

MONDAY — 2 OCT 44

All bents placed and braced, the first course of decking completed from the north abutment to bent 6 and second course to sill 2. Preparing to begin placing steel from south end. Job is 51% done. There is a critical shortage of 3" decking.

Personnel — Pvt Golden joined the Company, and Pvt Kolbusz rejoined from the hospital.

TUESDAY — 3 OCT 44

Steel erected span 6–8, being erected 8–9, 12–11 [*sic*]. Timber truss longitudinal bracing erected for the channel span (50'), 8–9. This bracing system is more complex than elsewhere in the bridge because of the length of the span. First course of deck complete to span 8. Job 65% done.

Lieutenant Deveikis was promoted from 2d to 1st Lt as of 1 Oct. All officers attended a Regimental officers' dinner with guests from Third Army Engineer Section and during the dinner the promotions were made, including the Captaincies of all Company Commanders. Present was Colonel Pearson, the first Regimental Commander, now with the Engineer's Office of 12th Army Group.

. . . .

FRIDAY — 6 OCT 44

The bridge is complete and was opened to traffic 1800 hours as estimated.

Personnel — Pvt Netski was dropped upon hospitalization.

SATURDAY — 7 OCT 44

Asphalting approaches to the bridge, loading surplus lumber to be used by Company "A," grading end of street connecting N57 at north end of bridge with bypass route to Treadway bridge at Liverdun, and opening additional waterway under span 1–2. Rip rapping with heavy stone the upstream ends of cribs and policing site.[8]

. . . .

BIVOUAC 17—TOUL, 1 MI WEST U6210, NORD DU GUERRE ZONE.

ARRIVED 1300 9 OCT 44—DEPARTED 1000 18 NOV 44.

MONDAY — 9 OCT 44

Arrived here at 1300 hrs, after a three hour trip, to build the biggest bridge to date. Reconnoitered the site over the River Moselle and the canal at Toul.[9]

Approximately 500' of two-way cl 40 bridge required for this crossing. At least six pile bents will be needed, of which four will have to be driven with floating equipment in order to use 50' spans thereby reducing substructure construction to a minimum. Suggested design contemplates the use of deep (24") I beams 50' long. Abutment bents can be driven with T/E [organic to the unit] equipment.

. . . .

TUESDAY — 10 OCT 44

Construction of the Toul bridge has begun. Cleared both banks of stream, cutting piling and hauling timber and small amount of decking. Decking will be a critical item on this project. Sketch is held up pending information on steel.

Personnel — Pvt Abels was hospitalized.

WEDNESDAY — 11 OCT 44

Toul bridge — Survey party established and located canal portion of the bridge. Preparing abutments for river portion of the bridge. Cutting and hauling piling, assembling timber and moving steel from Toul to bridge site. The signal wire on both banks of the canal must be removed.

Lieutenant Geesey with five enlisted men and acetylene burning equipment departed for "Werk Hagendingen" branch of the "Reichswerke Herman Goering" steel plant in the vicinity of Metz at 1400 to check possibility of moving 24" I beams from there and to begin cutting the steel to length if it is accessible. It is not possible to estimate completion date or to submit design of bridge until complete information is on hand on steel. . . .

Personnel — Tec 5 Nance and Pvt Abels were dropped from the rolls today, hospitalized previously.

THURSDAY — 12 OCT 44

Toul bridge — Stone-loaded crib was placed for the canal crossing. One abutment sill placed, the second being set. Assembling piling, steel, and timber at site.

Assembled a six-boat raft for floating drivers and began rigging the two pile drivers. The floating equipage is borrowed from the 1134th Engineer Combat Group.[10]

The steel detail returned late during the night, having been denied permission by the Engineer, XX Corps, to remove any steel from Hagendingen. The Colonel was notified. . . .

Personnel — Dropped Pvt Engler from the rolls.

FRIDAY — 13 OCT 44

Toul bridge — The steel is available at Hagendingen, as Lt Geesey reported last night. This morning the Colonel got clearance from Army Engineer and in the afternoon Lts Deveikis and Geesey with a detail of cutters and a fleet of 87th Engineer Heavy Ponton Battalion's trailers departed for Hagendingen again to haul 46 I beams 24" × 52'. The steel mill is an infantry Command Post. Fighting continues in its immediate vicinity and mortar shells are falling in the yards. The steel, however, has not been hit and it is deemed safe enough to work there with care.

. . . .

SATURDAY — 14 OCT 44

Toul bridge[11] — Completed driving bent 5, drove six piles for bent 6.[12] . . . Began setting steel span 1–2. Bulldozing east approach. Continue cutting and hauling piling. Although no word has been received from the steel detail, the first loads are expected tomorrow morning, 16 I beams. . . .

SUNDAY — 15 OCT 44

. . . .

Lieutenant Deveikis brought in three trailers of steel, reporting that one load was lost when the trailer skidded and wrecked on a muddy road in the artillery emplacements at Hagendingen when it took a wrong turn leaving the steel mill. . . .

Personnel — Tec 4 Woodward was reduced to private at his own request to be transferred out of the kitchen where he was not getting along with the Mess Sgt. Tec 5 Chalom was reduced for inefficiency as the 1st Platoon tool Cpl. Tec 4 Russell (promoted from Tec 5) replaced Woodward, Pfc Turner was promoted to replace Russell, and Tec 5 (promoted) Roush assumed Chalom's duties.

MONDAY — 16 OCT 44

Toul bridge — Completed driving bent 7. . . . The design sketch should be corrected to eliminate span 3–4.[13] This change was made today. . . . Eight loads of steel came in today; five more are due.

Tomorrow we will begin moving earth: approximately 6,000 cubic yds of fill are required for the approach road. The job is 11% done.[14]

The completion date of 31 Oct depends upon availability of decking in sufficient quantity to meet requirements previously reported. There is no decking in this area; all three company commanders have searched out the entire area. Even on the bridges at Pompey and St Vincent decking was

Inscribed photograph of Gen. George S. Patton. Courtesy Earl E. Hall

hauled from as far back as St Dizier. We have taken all 3" lumber from Army Engineer Supply Point [AESP] 8. We must be advised where [else] we can get this lumber. Nothing else is needed for the bridge except about 500T of premixed bituminous concrete for approach roads.

Personnel — Pvts Gainey and White received the Purple Heart for wounds received at Pompey. Promoted to Pfc: Abels, Westphal, and Hetler. Reduced — from Tec 5 to Pvt — Willmeth, from Pfc to Pvt Edmunds, Gryder, Strobel, and Zemla.

. . . .

WEDNESDAY — 18 OCT 44

Toul bridge — The river is up five feet during the night. The framing crews were unable to work today . . . or to continue driving piling except for one pile in bent 8 for a total of 105 to date. Progress of job will be directly affected by the high water.

. . . .

Began hauling earth fill to east approach road. We were depending on drag line[15] of 840th Engineer Aviation Battalion to load fill. At 1600 hours both machines left the site. In addition we have orders to return the B-E 15 crane to the 609th Engineer Light Equipment Company on Oct 20th. This rig is needed to load trucks and loss of it will delay completion about three days. The job is 16% done.

We are still uncertain about decking.

THURSDAY — 19 OCT 44

Toul bridge — The river is down slightly but work in the water is still extremely slow and difficult because of the swift current of the risen water.

. . . .

Installing leads and rigging drag line bucket on B-E 15 crane borrowed from 1301st Engineer GS Regiment. Began loading trucks again with fill for east approach road. The job is 25% complete. . . .

FRIDAY — 20 OCT 44

Toul Bridge — Bridging operations are suspended because of high water: the river is up six feet. Cannot cut off the piling in bents 6, 7, 8, and 9 until the river is below the elevation of caps.

Hauled fill to east approach until 1400 hours when the drag line had to be released to the 1301st. We are now rigging the 632d rig for this work.

The entire schedule and completion date is off keel because of the floods. We are behind three days already. Job is 25% done. . . .

SATURDAY — 21 OCT 44

Toul bridge — Operations on the river continue suspended. At its highest stage to date at noon, the river has begun to fall. The hauling of fill to east approach continued all day.

Roads — Demudding continues unchanged.

A special report on the condition of the bridges within the company area was rendered today.

SUNDAY — 22 OCT 44

Toul bridge — The river has continued to fall but it is still too high to permit operations. . . .

MONDAY — 23 OCT 44

Toul bridge — Began spacing piles in bents 7 and 8 for capping. . . . Job 26% complete. . . .

. . . .

WEDNESDAY — 25 OCT 44

Toul bridge — . . . Setting forms for concreting west approach. The drag line was broken down most of the day so no fill was hauled [to] east approach, and there was no Quickway available for loading back-fill material for widening the connecting road on the west approach. Obtained concrete mixer at Commercy;[16] it has been repaired and is in operating condition. Job 33% done.

To meet our needs for base and fill material we have located a quarry on N4 three miles west Toul where a shovel is loading for 293d Engineers. The total haul for us would be four miles. Recommend that Battalion place a shovel there tomorrow to furnish the remainder of our material for the approach roads and for improving water points which we are maintaining. . . .

Personnel — Pfc Keller went to the hospital and was dropped from the rolls.

THURSDAY — 26 OCT 44

Toul bridge — Completed framing bent 10. . . . Moving the floating rig to position to raise timber bent at 6 and to erect steel span 5–6.

Setting forms and subgrading connecting road, west. Back-filling over anchorages of abutment bulkheads at 10. Job 40% done. . . .

The 390th Engineer General Service Regiment assumed maintenance and security responsibility for the railroad bridge over the Moselle at Toul, relieving us.

. . . .

SUNDAY — 29 OCT 44

Toul bridge — Steel set span 6–7, set timber bent 8, finished decking span 5–6. Filling east approach and concreting West approach. Job 60% done.

The new completion date is estimated at 5 Nov, extended beyond the original estimate to compensate for the delay caused by flooded river. . . .

Roads—Demudding at Quartermaster Supply Points I and II on N4 west of Toul completed to the satisfaction of the officer in charge except for the drainage ditch. This ditch cannot be opened by bulldozer or grader as equipment bogged down before it even reached the location. Nor can the ditch be shoveled out by hand, the mud being of about the consistency of butterscotch pudding. To drain the area immediately it seems best to use a drag line to dig [a] wide-bottom ditch with flat slopes.

MONDAY—30 OCT 44

. . . .

Personnel—New men assigned: Tec 5 Mataitis and Pvt Kreeft. Promotions: to Tec 5 Howery, Joseph Kelly, and Plucinnik. Hospitalized and dropped from rolls: Tec 5 Howery, Jaspersohn, and Pasquini. [Jasperson rejoined the unit on 2 November, Howery on 4 November, and Pasquini on 7 November.]

TUESDAY—31 OCT 44

. . . .

Island paving—No concrete poured today. Forms are limited in number, and the concrete poured to date is too green to permit stripping its forms. Completion remains at 85%.

Roads—A tractor and grader are 98% complete draining the Quartermaster supply point.

Personnel—Lieutenant Geesey was relieved of assignment to this company . . . and reassigned to Company "D." . . . He departed today. . . .

THURSDAY—2 NOV 44

Toul bridge—. . . Began laying selected material surface course on east approach, and continue concreting west approach. Bridge 80% done.

. . . .

Roads—Demudding has been increased to include scraping of N404 and Ic 11 in addition to N4 and maintenance of roads for water dp [distribution point] 44B.

. . . .

FRIDAY—3 NOV 44

Toul bridge—All steel is set and the first decking is complete. The second course of deck is being put on spans 7–8-and 8–9. Constructing walks and handrails.

The earth fill on the east approach is complete, as is the concreting of the west approach. Bridge is 98% done.

Island paving — complete 100%.

Because of the damp weather and rain, coarse-ground cement and clay in the aggregate, the pavement in the west approach and in the island is unusually slow in setting. It was recommended not to open these lanes to traffic till the 12th.

Roads — demudding always goes on.

SATURDAY — 4 NOV 44

Toul bridge — Completed the second course of decking, walks, and railings ending all construction work on the bridge. There remains some cleanup work and dressing of the shoulders on the approaches. This is the final report on the Toul bridge.

Fender piling — Moved the floating rig into position and started driving piling for the center pile on the railroad bridge over the Moselle. Job 75% complete.

Roads — Demudding, of course.

. . . .

WORK ON THE TOUL BRIDGE included "approximately 500 sq yds of concrete paving poured on the west approach and approximately 1000' of road was widened and improved on the east approach." The bridge did not open until 6 November, and on the night of 8–9

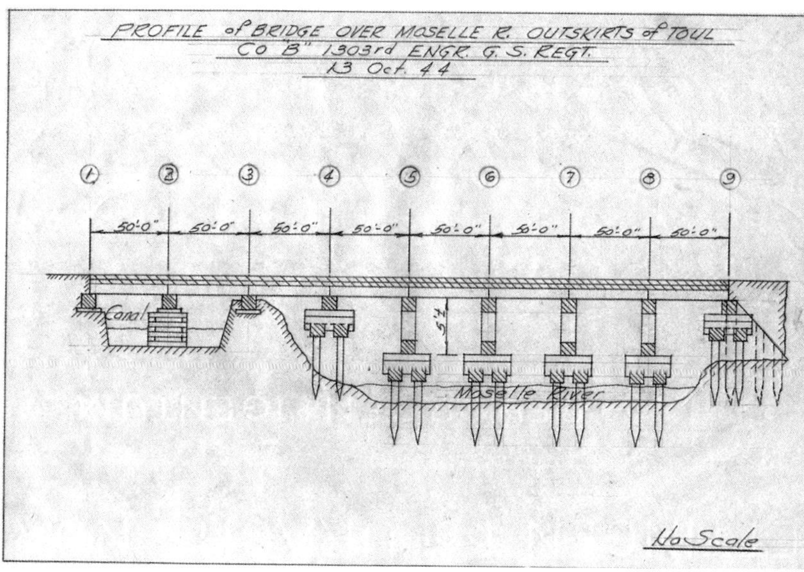

Drawing of the Toul bridge. Courtesy Earl E. Hall

Decking the Toul bridge. Courtesy Dwight Gowdey

Toul bridge completed (roadway view). Courtesy Dwight Dowdey

Toul bridge completed (side view). Courtesy Dwight Gowdey

November, flood waters rose 3 inches above the bottom of the steel stringers, threatening to wash out the bridge. The water began receding, and the bridge survived what locals called "the maximum all-time flood."[17]

SUNDAY — 5 NOV 44

Fender piling — Completed driving piling for the fender for Moselle River railroad bridge at Toul.

Started positioning the floating rig to drive fender piling for the new highway bridge (our bridge). The crane has to be taken off the raft as the complete rig will not clear the bridge. Completion estimated 8 Nov: 2% done.

Policing of Toul bridge site continues: building shoulders on west approach; placing guard rail post at east approach; placing the shack for bridge guard at west end of bridge; removing excess lumber from site to Regimental dump.

Roads — Demudding continues unchanged.

Personnel — Tec 5 [Hughie] Magee, while on detail hauling excess lumber from Toul to Pont St Vincent, fell from a moving 20 ton trailer at 1730 hours and was seriously injured. He was taken to a hospital and dropped from rolls. [Magee "died in the hospital" on 6 November 1944.]

MONDAY — 6 NOV 44

. . . .

Policing of Toul bridge site continues: setting guard rail, graveling foot path at west approach. Since higher headquarters disregarded the recommendation to allow the concrete to set up properly, the Toul bridge was opened to traffic at 1200 hours as ordered.

. . . .

Bridge guard — Relieved 178th Engineers of maintenance guard on tactical bridge at Liverdun 0800 hrs,[18] the security for which is being provided by 6th Cavalry Reconnaissance Group. Meanwhile, preparatory to assuming new responsibilities, relieved of all bridge guard duty in Toul.

. . . .

TUESDAY — 7 NOV 44

Fender piling — Driver deadlined for repairs. . . .

Roads — Using dry cement to patch N411: Opening ditches. Bridge guard — In addition to continuing the maintenance guard on the Liverdun floating bridge, assumed responsibility for tactical guard on the following bridges: (1) Bailey Bridge over canal at Liverdun; (2) Timber Trestle over canal at Liverdun; (3) Canal flume bridge over river at Liverdun; (4) pile bent

A crowd watches as soldiers check the Troyes bridge. Courtesy Dwight Gowdey

bridge over river and canal at Toul; (5) Timber Trestle (former swing span) on N404 at Toul; (6) SA [reference unclear] at road junction [RJ] N4-N408 at Toul.

. . . .

WEDNESDAY — 8 NOV 44

Fender piling — Suspended operations which were never begun. The raft has been dismantled and all balk and chess is stacked above high water mark. Pontons are tied up in the quiet water.

Policing of the Toul bridge site is complete.

Roads — Opened drainage ditches on N411 but the weather is too wet for further patching. The rains continue and the river is rising.

Bridge guard — Continue guarding bridges (1), (2), and (3) using four NCOs and 24 enlisted men.

Bulldozers were sent to Liverdun at 1300 hours to dam off the flow of water into the canal flume.

At 2000 hours the entire Company — less bridge guards, details, and power boat operators school personnel — fell out to protect our Toul bridge from damage by the rising river and floating debris. The upstream walk was torn up so that debris wouldn't bank up against the I beams and under the

Flood waters rising at the Toul bridge. Courtesy Dwight Gowdey

walk. Rock rip rap was placed around the toe of the fill on the east approach. Coffee and sandwiches were sent out at midnight. The water was only several inches short of the steel.

Personnel — Lieutenant McKNIGHT [who "was assigned to and joined the company" on 2 November] was happy to get his assignment to the 35th Infantry Division today. Relieved of assignment to this Company, he did not depart today.

HEAVY RAINS ON 6–8 November "caused a very rapid rise in the Moselle and the river reached a flood stage which had not been equaled in 29 years. . . . During the period 8, 9, and 10 Nov[ember,] readings were taken hourly on gauges at Toul and Pont St. Vincent and Third U.S. Army headquarters was constantly informed of the river stage. At Toul the river rose rapidly until 0130 hours 9 Nov. From 1600 hours 8 Nov to 0130 hours 9 Nov the rise was approximately three feet. . . . From 0130 hours to 0600 hours 9 Nov the rise was four inches. By 13 Nov the river was back to normal."[19] Company B was at the bridge, working to clear debris, "illuminated by headlights of two trucks," and took whatever safety precautions were possible. The men removed the sidewalk on the upstream side during the night to facilitate this task. The water rose approximately three inches above the bottom of the steel stringers, and it seemed as if the bridge would soon be washed away. But the flood had crested. Residents reported that this had been the highest water in twenty-nine years. The only damage to the bridge was some "slight settlement and some erosion of the new fill on the east approach."[20]

FRIDAY — 10 NOV 44

. . . Bridge guard — Continues on bridges 1, 2, and 3. The flume over the canal bridge remains flooded.

. . . .

Toul bridge — Maintenance guard continues on bridge keeping debris clear of the bridge. The washout in the east approach fill is being repaired. Began replacing the rail on upstream side. Reopened to traffic 2000 hrs.

Personnel — Pfc Abels was reassigned to the Company. Promoted to Pfc were Biarruel, Clary, Edmunds, Farr, Gryder, Olson, Rosinski, and Zizzi.

SATURDAY — 11 NOV 44

Fender piling — Operations will resume tomorrow. Estimate completion for the 15th. Project prior to suspension 5% done.

Roads — Patched chug holes in N411 with dry concrete. This method is satisfactory if the new patches are not subjected to hard rain and heavy traffic too soon. At 1800 hours all responsibility for N411 and all roads ceases.

Toul bridge — Completed replacing guard rail upstream side and repairing east approach. The 25 Ton pontons were removed from the water and cleaned prior to returning them to 1134th Group.

Bridge guard — Continue guarding bridges 1, 2, and 3. At 1800 hours security guards were posted on bridge 46, railroad underpass on N411, and the three bridges on N4.

Personnel — Lt McKnight departed for 35th Division today, and Lt Howard was assigned to the Company. Prior to his injury while riding a motorcycle in England and hospitalization, Lt HOWARD was Headquarters and Service Company Commander, this Regiment. [On 4 December 1944, "Lt Howard was relieved of assignment to the Company and reassigned to Company 'E.' The 2d Platoon lost another officer."]

SUNDAY — 12 NOV 44

Fender piling — Relieved of this assignment. The rig is being changed from a pile driver to a shovel.

. . . .

MONDAY — 13 NOV 44

. . . We drained the canal flume bridge, No. 42, and reopened it to traffic.

Readings — 2.85, 2.80, 2.70.

. . . .

WEDNESDAY — 15 NOV 44

. . . Made minor repairs to approaches of Toul bridge.

Fender piling, Pompey — Moved crane to bridge No. 17 at Pompey and started rigging leads for driving fender piling.

Readings — 2.40, 2.35, 2.30.

Reconnoitering Nancy for billets. Because Third Army Hq is in that town, billets are extremely difficult to get as Army does not want any except their own headquarters troops in town[,] to make it a boisterous place[,] or an eyesore to the French people, whom everyone is going to great lengths to please and impress favorably.

THURSDAY — 16 NOV 44

. . . Returned all ponton equipment to 87th Engineer Heavy Ponton Battalion.

Readings — 0.84, 0.81, 0.78.

Personnel — Pfc Farr [hospitalized on 14 November "and dropped from the rolls"] was reassigned to the Company.

The billets in Nancy were finally approved. Accompanying the approval was this letter. From the Office of the Regimental Commander the subject—"Conduct of Troops in Nancy"—was handled like this: "1. In order to obtain the approval for Company B billets in Nancy it was necessary for the Army Engineer to explain fully to the Chief of Staff the importance of having the troops in the city. The Chief of Staff finally approved the billets after they had been disapproved by the Asst. Chief of Staff. Even so, Army still does not want surplus troops in Nancy and only in rare instances will approval for billets be given. 2. After the Army Engineer had personally obtained approval of the B Company billets in Nancy, he then explained fully to the undersigned the importance of following to the utmost all rules and regulations pertaining to troops in Nancy. 3. You will impress in the mind of every man in your unit the utmost importance of obeying all rules and regulations for troops in Nancy. All the military courtesy and discipline that has been taught to your men will be strictly practiced. If you have a guard in front of your headquarters he will salute every officer whether the officer is walking or riding. Your men will be kept out of beer joints and in all respects properly controlled." It was signed by the Colonel. And, of course, Major Mylius added his indorsement [*sic*]. "For strict compliance," he said.

. . . .

BIVOUAC 18—NANCY, U8509, NORD DU GUERRE ZONE.

ARRIVED 1200 18 NOV 44—DEPARTED 1000 9 DEC 44.

SATURDAY—18 NOV 44

Opened Command Post in Building B, Quartier Druot, Rommel Kaserne, Nancy, at 1200 hrs, after a two hour trip from the old area.

Bridge guard—Continues unchanged, on bridges 21, 24, and 26.

. . . .

SUNDAY—19 NOV 44

. . . .

Bridges—Completed thrust bracing bridge 24 100%. The placing and grouting of rip rap counterfill will begin tomorrow and will be considered a separate job. Estimate completion on 28 Nov.

Ice breakers—Assembling steel and breaking holes in grouted rip rap on bridge 24. . . .

The following floating equipage was assembled and is being used at Pompey: 20 pontons, 142 ponton balk, 36 trestle balk, 170 chess, 4 half chess, 54

C clamps, 3 jacks, 6 bilge pumps, 11 oars, 2 boat hocks, 19 side rails, 1 assault boat, 6 paddles.

MONDAY — 20 NOV 44

. . . .

Bridge maintenance — Counterfill at bridge 24 started today with the moving of concrete mixers to site, hauling rubble for fill, and building chute for concrete. Job 2% done.

Ice breakers — Continue assembling steel, started rigging tackle for erection of steel. At noon the job was suspended and operations will not be resumed until arc welder and burning equipment are available. Completion unchanged: 2%.

Readings — Toul gauge was not read today. The Nancy bridge reading was phoned in to the Army Engineer at 0700 hrs: 0.5.

. . . .

WEDNESDAY — 22 NOV 44

. . . .

Fender piling, Pompey — No work was done today because of the rising river. The crew worked until midnight last night securing equipment on the island. Completion remains at 39%. . . .

Hauled bridge timbers from Toul to Pont St Vincent dump. Readings — Toul 1.5, 2.4, 1.8; Nancy 1.0, 1.2, 1.7.

THURSDAY — 23 NOV 44

. . . .

Fender piling, Pompey — Again it was not possible to work because of high water. Completion will be delayed as many days as high water obtains. . . .

Materials — Found about eighty 50' rolls of barbed wire in abandoned stables in billet area. Recommend Regimental Supply Officer remove them to Engineer dump.

Readings — Toul 1.6, 1.0, 1.0; Nancy 2.1, 2.5, 2.3.

. . . .

SATURDAY — 25 NOV 44

. . . .

Readings — Toul 2.3, 2.4, 2.5; Nancy 4.0, 5.2, 5.4.

. . . .

Bridges — Counterfill job suspended: men being used on bridge maintenance, guarding against destruction by flood waters the existing bridges. The incessant heavy rains are troublesome.

Last night the high water tore out four spans of our Toul bridge. One span from the west bank is still in place but the off shore bent support is apparently damaged because of the length of this span. Only a bridge guard was present and he could do nothing either to save or salvage the bridge.

It is not recommended to replace this fixed bridge with a Bailey, especially not a floating Bailey because of the difficult[y] that would be met anchoring boats against the rapid current and because of the extreme differences between high and low water levels in the river. If this site must be bridged, recommended a pile bent bridge with a greater vertical clearance, a higher bridge, simply.

Readings — Toul 2.3, 2.4, 2.5; Nancy 5.1, 5.0, 5.2.

The bridge guard in Nancy questioned three men dressed as French civilians when they behaved suspiciously on a bridge. Unable to produce proper French identification, they were turned over to the Military Police.

Personnel — Pvt Rosinski in arrest in quarters awaiting Court-Martial for violations of the 61st and 96th Articles of War.[21]

Toul bridge destroyed in flood. Courtesy Dwight Gowdey

Decking of Troyes bridge in floodwaters. Courtesy Dwight Gowdey

DURING THIS FLOOD, water was approximately one foot higher than the level of the same time in 1943 and "about 1100 hours, a section of the bridge broke away. Subsequently other sections collapsed, and by 1400 hours, only 100 feet of the original 300 feet of bridge remained. The embankment at the east end was not damaged. According to eye-witnesses the bridge first failed at the first pier from the east bank. The deck on the upstream side appeared to drop and then the pier apparently went out. The drag of the debris probably resulted in the failure of the other spans. . . . Most of the deck that was on the demolished portion of the bridge was in two pieces which had floated downstream." Some steel stringer remained in place, but others were carried downstream. The decking was so well nailed together that it was difficult to salvage. Since it was no longer essential to the tactical situation, the bridge was not replaced. Company A took on the salvage work as Company B was given a new assignment.[22]

MONDAY — 27 NOV 44

. . . .

Counterfill job and fender piling tasks were suspended indefinitely on orders from Regimental Headquarters. Bridge removal has priority [*sic*].

Bridge removal — Assigned the removal of Bailey bridges at Custines (15) and Champigneulles (20). No. 15 is a DD/DT 250' long. Removal started

1500 hrs. Work will continue unbroken. Expect to be ready to load parts at noon tomorrow. No. 20 is an 80' D/S. Its removal was also started at 1300 hrs.

Readings—Toul 2.0, 1.95, 1.85; Nancy 5.8, 5.6, 4.4.

Personnel—Pvt Gruppi was hospitalized and dropped from strength. [On 29 November "Pvt Gruppi [was] reassigned to and rejoined the company."] Pvt Rosinski released from arrest in quarters pending trial by Court-Martial.

TUESDAY—28 *NOV* 44

. . . .

Bridge removal. No. 20 at Champigneulles was dismantled and stockpiled by 0200 hours this morning, and hauled away at 1400 hours this afternoon. The Custines bridge (No. 15) required launching nose and extra tools which were hauled to site. Dismantling started late and by 1400 hours this afternoon the bridge was up on 14 rocking rollers. Because of a lack of space on the east end of the bridge it is being dismantled 80' at a time and loaded onto trucks directly. This work was interrupted this afternoon to release men for loading the Champigneulles Bailey.

Salvage—A small party rafted and secured usable debris from the wrecked Toul bridge between Toul and Gondreville on the river Moselle.

Readings—Toul 1.55, 1.22, 1.14; Nancy 3.7, 3.4, 3.0.

. . . .

THURSDAY—30 *NOV* 44

. . . .

Bridge removal, Custines.—Dismantling was completed during the night and ten trucks loaded. Four trucks became available at 1300 and were loaded. At 1330 28 trucks arrived, the rest of the bridge was loaded into ten of them and the remainder plus the crane were returned.

. . . .

Readings—Toul 1.0, 1.1, 1.0; Nancy 2.0, 1.0, 1.6.

FRIDAY—1 *DEC* 44

. . . .

Bridge maintenance—a. Fender piling, Pompey. Resumed operations at 1600, after rigging leads completely and dismantling the raft in the small stream. The job is 45% done; new completion date estimated 3 Dec.

In addition we rebuilt the floating bridge over the small stream in order that the 1134th Group might have access to the island to remove their equipage.

b. Counterfill, Nancy. Operations resumed this morning. The work previously done was damaged to the extent that completion had to be revalued at 50% and the final date advanced to 7 Dec to compensate for the four days lost to the flood.

. . . .

Personnel—As the culmination of a long series of temperamental differences between his superior noncommissioned officers and himself, Cpl [Harry] Huntington was reduced to Private at his own request.[23] Shouse replaced him and Chavez took the former post.

SATURDAY—2 DEC 44

. . . .

Personnel—Pvt Hughes,[24] on Temporary Duty at Regimental Headquarters as a guard, while driving with a Regimental Headquarters courier in a jeep, was injured last night. Stopping for a break, he retired to an alley and came out following a gunshot with a shoulder wound. The indicated assailant could not be traced in the blind-alley, a dead-end passage. Hospitalized, absent sick. LD undetermined [was this a Line of Duty injury?]: the Colonel has ordered an investigation.

. . . .

MONDAY—4 DEC 44

. . . The 1134th Engineer Combat Group continues removing equipage and loading steel at the site for our Regimental dump.

. . . .

Miscellaneous—a. Built four latrines for 35th Evacuation Hospital. b. Reconnoitered new site for 65th Field Hospital. No Engineer work required. Cleaning being done by medical personnel.

. . . .

Special Bridge Report. a. Foundations. At bridges 22, 24, 28, and 46 the foundations were checked and rechecked over a period of five days. No deterioration is indicated. On 24 the new rip rap has scoured away on the upstream face of the outboard pier in the remaining structure on the east bank. This cavity is being filled.

. . . .

Personnel—. . . Pvt Aiken was hospitalized and dropped from the rolls.

. . . .

WEDNESDAY—6 DEC 44

. . . .

Miscellaneous—Continue helping 1134th to clear out their floating equipage. Cleaning equipment at bridge 24. The 35th Evacuation Hospital picked up their latrine boxes today.

Deicing—Located unlimited supply of deicing materials in Pompey Steel Mill on old Custines bridge road. Hauling to N57 starts tomorrow. Also, reported to Regimental Headquarters that mounting the snow plow on 2–1/2T truck will damage king pins, bushings, and spindle, and cause the truck to shimmy.

Readings—Toul 1.6, 1.4, 1.2; Nancy 1.4, 1.6, 2.0.

Personnel—Cpl Brehl . . . was replaced on Temporary Duty at Regimental Headquarters as Company Clerk by Tec 5 Groom.[25] [The transfer was effective 10 December.]

THURSDAY—7 DEC 44

. . . .

Deicing—Loading anti-skid material by hand at Pompey and stockpiling along N57.

. . . .

Personnel—Pfc Cederberg reassigned to and rejoined Company. [He had been "hospitalized, dropped from rolls" on 3 December.]

FRIDAY—8 DEC 44

. . . .

Personnel—Lt Everett, commissioned from Tec 5 in Company "A," was assigned to and joined Company today, primarily for orientation for his new duties under the tutelage of Lt Deveikis.

BIVOUAC 19—NANCY, 1 MI E, U9013, NORD DU GUERRE ZONE.

ARRIVED 1100 9 DEC 44—DEPARTED 0730 21 DEC 44.

SATURDAY—9 DEC 44

Hospitals being what they are, there was not much use trying to hold the good billets for ourselves when a hospital wanted to move to Nancy. We departed at 1000 and located in the Kaserne Kleber at St Max on the outskirts of Nancy one hour later.

. . . .

SUNDAY — 10 DEC 44

. . . .

Reconnoitered the road system of Limey and Andilly military cemeteries preparatory to assuming maintenance work. [Rain delayed grading work until 12 December.]

To maintain open roads during the winter, a survey is being made among civilian highway maintenance foremen who will work their crews spreading anti-skid material when required.

. . . .

WEDNESDAY — 13 DEC 44

. . . .

After three attempts we still cannot find the gauge on the Millery bridge. Requested advice from higher headquarters after we could not find the gauge on the locks near Marbache where Army Engineer advised us to look.

Personnel — Pfc Rosinski reduced to Pvt by CMO [court-martial order] 27.

THURSDAY — 14 DEC 44

. . . .

We are assisting CWS [Chemical Weapons Service] Depot No. 7 to move from Flirey.

Submitted today special report on French civilian road maintenance organization.[26]

. . . Personnel — Pvt Kreeft replaced Strobel as Regimental Headquarters guard on Temporary Duty. [On 5 December "Pvt Strobel [was] placed on Temporary Duty as Regimental Headquarters guard."]

FRIDAY — 15 DEC 44

. . . .

Continue assisting Chemical Warfare Service depot 7 to move. The heavy equipment borrowed from Company A is being returned because the 633d Engineer Light Equipment Company is being detached from the Regiment. Replacement of this equipment is vital as three tractors and one crane are required for three or four more days.

. . . The 676th Engineer Light Equipment Company relieved the 633d and the 1st Platoon of the new company is assigned to the 1st Battalion.

SATURDAY — 16 DEC 44

. . . Impossible to estimate completion date of work moving the Chemical Warfare Service depot out of the mud.

Readings — Toul 0.4, 0.4, 0.4; Nancy 1.0, 1.2, 1.2. Because we cannot located the reported Millery gauge, we will install a new one on orders of S-3.

Third Army is pushing forward on both sides of Saarbrucken. The Seventh Army has crossed the German frontier [*sic*] and is shelling the Saar Valley city of Karlsruhe. In the Pacific another landing in the Philippines is reported, on Mindanao.

SUNDAY — 17 DEC 44

. . . .

Continue moving CWS depot at Flirey. We are pulling their heavy equipment out of the mud onto hard roads. Reconnoitered railroad sidings available at Pompey and Pont-a-Mousson.

Readings — Toul 0.3, 0.25, 0.2; Nancy 1.1, 1.0, 1.0. We are building the Millery gauge: it will be placed tomorrow.

The Military Police Detachment Commander, Chateau Salins, reported today that the Bailey Bridge at Puttingny, on Gc131, is not needed as no traffic is using it. Recommended to Army Engineer removal of the bridge.

Personnel — Tec 4 Sucher took Mataitis, Clark, Storm, Plucinnik, and Jenkins with him on Temporary Duty to 1134 Engineer Combat Group to do carpentry work repairing the floating equipage for Third Army, vicinity Toul, where Colonel Fitch's dump is located. Departed 0900.

The Third Army is in Germany now on a 40-mile front, as is the Seventh, for a total of four American armies there. Last night German planes were overhead for three to four hours. This was their first appearance in many weeks.

MONDAY — 18 DEC 44

. . . .

1st Battalion Headquarters messaged us as follows: "Unconfirmed report (from Army) that parachutists have been dropped along N74 about five kilometers south of Nancy between 2000 and 2200 hours 17 Dec. Advise any of your troops in this area to take necessary precautions." It caused a stir when received: all day everyone's talked of German paratroopers.

Not without reason. The Germans are making a determined counterattack in the Ardennes Forest on the First Army front, adjacent to our own (the Third). They have dropped men by parachute and are using several Panzer divisions and several infantry divisions in their attempt to push back the Americans on a 60-mile front. Airborne troops are disrupting our communications. Scattered paratroopers are reported dropped in Third Army area. The last report has it the attack has been contained.

Personnel — Cpl Armstrong was honorably discharged [from] the Service to accept commission as 2d Lt, CE, today, and this evening the Company officers entertained him at the Command Post. He is, unfortunately, being transferred: 150th Engineer Combat Battalion.

Cpl Brehl was reduced to grade of Private and Tec 5 Groom promoted to Cpl. Pfc DeMattei and Pvt Willmeth were upped to Tecs 5.

TUESDAY — 19 DEC 44

Bridge guard continues unchanged in Nancy. We added two at Pompey, 17 and 18.

Road maintenance continues unchanged.

Deicing — stockpiling of material on N4 continues. In addition, anti-skid material is being loaded on gondola cars at Pompey for rail shipment.

Continue moving the Chemical Warfare Service depot at Flirey.

. . . Reports we have received indicate that the German offensive has been contained by the inadequate American forces who were caught literally with their pants down. The Luftwaffe was up in strength for large air battles. The Battalions of this Regiment are alerted for movement forward to relieve the 1137th Engineer Combat Group who might be called forward to help halt the Germans south of Aachen.

WEDNESDAY — 20 DEC 44

We were relieved of responsibility for bridge guard on bridges 17, 18, 24, 25, and 26 at 1200 hrs; all road maintenance moving Chemical Warfare Service Depot 7 at Flirey and of reading the river water level gauges. The final readings herewith: Toul 0.0, 0.0; Nancy — gauge found broken this morning; Millery 1.5, 1.5.

The Regiment is ordered into Luxembourg: Company B to vicinity of Esch. XII Corps is being shifted because of the German breakthrough. The reports of "containing" the Germans are somewhat ridiculous: they have come up to within 20 miles in Luxembourg.[27]

To get a head start on the hunt for billets, Lt Deveikis took four men from his 2d Platoon and with the weapons carrier departed at 2245 this evening from France as billeting officer.

Personnel — Pvts Woodward and Isham Wilson were marked AWOL at 2000, failing to report back from pass.

CHAPTER SEVEN

Luxembourg, on the Flank of the Battle of the Bulge

DECEMBER 1944–MARCH 1945

In the fall and winter of 1944–45, Allied armies were at the end of a very long and thin logistical line. Operation Dragoon's opening of Marseilles and the use of French railroads in the south and east helped compensate for the devastation of northern France's transportation network. Allied air power, helped by the French Resistance, had so destroyed the railroads, bridges, and other infrastructure that troops were functioning in what might be called a "supply desert." Supplies, especially of gasoline, remained tight, and worse yet, manpower issues now became a problem. Gen. George Marshall's insistence on the "90 division" strategy meant that the U.S. Army had too few infantry replacements trained and ready as casualties mounted. Most "noncombat" units in Europe, like the 1303rd, were taxed to provide young, fit men for use as infantry; since the 1303rd was a combat-zone unit, it was exempted from this order. Consequently, units on the line would stay on the line, without relief for rest and reinforcements. In addition, the Allies seriously underestimated the Germans' ability to replenish their own military forces.[1]

These problems came together during the German Ardennes offensive, as the "Battle of the Bulge" is properly known. This counterattack posed serious dangers to the Allied advance and to the integrity of the Allied XXI Army Group and the U.S. First Army. Fortunately, Patton's G-2 (intelligence officer), Col. Oscar W. Koch, had great skill in tracking German and Allied movements throughout northern Europe. As a result, he forecast a possible German assault in the Ardennes region, specifically against the First Army. Prepared with contingency plans, Patton was able to make a brilliant movement of fighting divisions and logistical support units, including the 1303rd EGSR, northward into position to open, on 26 December 1944, a supply corridor to the surrounded paratroopers of the 101st Airborne Division at Bastogne. This shift of the axis of advance, done largely on verbal orders, was unusual in its daring and complexity.[2]

Among the units shifting 90 degrees and moving at their best speed across an icy, snow-covered landscape was the 1303rd. Ordered into Luxembourg, the engineers again were on the right flank of the XII Corps and Third Army. This time the danger was more acute since weather often grounded any air support. Arriving in Luxembourg, the first order of business was to lay out defensive posi-

The 1303rd's Service in Luxembourg. Courtesy Paul A. Lorah

tions. The 1303rd was responsible for protecting miles on the flank, far more than a division normally should cover. Company B found itself sent to a crossroads, where a self-propelled artillery battery was posted. As the company — unsupported on its flanks — laid out its small line of positions, the artillerymen packed up and moved to a more defensible position. They had no faith in the fighting abilities of the engineers.[3]

LUXEMBOURG

BIVOUAC 20 — DUDELANGE, U8199, NORD DU GUERRE ZONE.

ARRIVED 1730 21 DEC 44 — DEPARTED 1030 12 JAN 45.

THURSDAY — 21 DEC 44

The billeting party crossed the French-Luxembourg border at 0200 this morning; arrived at Esch 0230; slept in their vehicle in the square until daylight; got breakfast at the MP mess; and began billet reconnaissance. After considerable difficulty and finding no billets not occupied by division troops and their huge supply stocks, particularly the 80th Division, with the aid of Civil Affairs, the Company was assigned the Summer school for tubercular children in the Lehwald on the outskirts of Düdelingen, near the border. A light snow fell during the night.

Meanwhile, the Company departed in convoy from the last bivouac in

France at 0700 and arrived at this first bivouac in Luxembourg at 1730, at dusk. Two dozers, for which no prime movers were available, will be brought up later.

Personnel — Woodward and Isham Wilson returned from AWOL to duty status as of 0800 this morning.

The Regiment is in Luxembourg now, with the 1st Battalion concentrated in Differdange and the Company in Dudelange.

FRIDAY — 22 DEC 44

We awoke in a snow-covered world and spent the day getting established in our new area. The evergreens with their snow mantles look like a yard-full of Christmas trees: lovely.

The Battalion Construction Foreman — S Sgt Stoudt — has been commissioned a 2d Lt, assigned to the 1301st, and has started a minor war as the Colonel wants his valuable construction knowledge.

Silence shrouds the German offensive. We are waiting.

SATURDAY — 23 DEC 44

Completed assigned road reconnaissance and reported.

Posted security guard on bridges at Hettange and Grande.

Grader clearing snow from assigned road net.

The Luftwaffe is very active. Yesterday 178 German planes were shot down in the clear weather.[4] Last night extremely heavy antiaircraft fire peppered the bombing planes which appeared again in the brilliant moonlight. The northern flank of the German offensive has been stabilized somewhat but the southern one continues fluid.

SUNDAY — 24 DEC 44

Bridge guard — Relieved of security guard on Hettange and Grande bridges. Posted security guard on main highway railroad bridge on N4 in Luxembourg City relieving Company "F" of this at 2000.

Road maintenance — a. Made initial survey for construction of road to bridge dump vicinity Esch, Army Engineer Supply Point No. 13. b. Reconnoitered Villerupt and Longwy area. Maintaining N53. Located deicing material and surveyed sites for stockpiles.

Personnel — Sgt Baker was honorably discharged to accept commission as 2d Lt. Tec 5 Jenkins (on Temporary Duty) promoted to Cpl and Pfc Cooper to Tec 5.

What might have been a very sad Christmas Eve was made bright by some good news. Third Army has hit the German southern flank, crossed most of Luxembourg, and is driving for the center of the German attack in Belgium. Clear weather made possible some 7,000 sorties yesterday. One force was made up of 2,000 Forts and Liberators with 900 fighter escorts. With Third Army in the fight, everyone can relax.

The Regiment is alerted to defend, if necessary, the southern Flank of Third Army. The defense plan calls for demolitions and dug-in defensive positions based on all available weapons: riflemen, machine guns, and bazookas as well as rifle grenadiers. Orders received late at night.

THE REGIMENT'S PLANNED DEFENSIVE line and barrier zone from Sandweiler to Berg, including strong points and positions, was created from maps. On 25 December all units made a reconnaissance of their sectors to finalize plans. "All available men were immediately set to work organizing these positions. Fox-holes were dug, weapon positions were selected and dug in, and road blocks and barriers were planned. . . . All suitable positions for demolitions (bridges, culverts, wooded areas, ravines, etc.) were reconnoitered and a complete demolition plan for the area was prepared. The charges were prepared, placed, then removed, and the explosives returned to company areas. . . . Command Posts for this defensive line were connected by telephone with each other and with higher headquarters. Forward units were contacted and defense plans coordinated with these units and with adjacent units." The 764th Engineer Dump Truck Company and 676th Engineer Light Equipment Company assisted. Earl Hall has noted that the deployment order put Company B in a position about a half-mile wide, centered on a road, and without any support on its flanks. A self-propelled artillery unit occupied the crossroads, but apparently suspicious of the quality of support the engineers could offer, the artillerymen moved elsewhere. The company's orders were to hold out and delay the Germans as long as possible. Since the position was untenable, the engineers were very pleased that no Germans chose to contest the crossroads.[5]

MONDAY — 25 DEC 44

Christmas Day we spent digging in. This is the first time since our arrival on the Continent. Everyone worked except the details and bridge guard. The emplacements are near the border, tied in with similar positions of other troops in the area: the scattering, a thin one, would be no more than a delaying action because, if it were put into operation, the outcome is obvious. The positions, not entirely complete, could be used as of 1800 tonight.

We did have a hot supper: Christmas turkey. It was excellent. Probably the locally obtained beer and white wine helped the mood. The weather is cold and clear, and the scenically beautiful countryside is a winter wonderland except where the spoil of fox holes lies like dark rings upon the snow.

At church services this evening the Chaplain passed out little greeting cards issued by General Patton.[6] They contain his greetings and his prayer for good weather which must have been heard as this is the fourth consecutive clear day. The air force has operated tactically with brilliance. The Germans are some 50 miles inside Belgium, their center of attack strong. Our flanks are attacking vigorously and with success: the Third Army on the south, the regrouped First on the north.

Personnel — Gavinsky replaced Fast on guard at Regimental Headquarters.

TUESDAY — 26 DEC 44

. . . .

Lt Hall departed with detail to haul bridging material forward from Pont St Vincent.

Defensive positions — Completing the improvement of the assigned positions for Army flank defense. Although physical cratering of roads has not been accomplished, the sites are marked and all plans made. Detailed report will be submitted.

Personnel — Lt Everett was relieved of assignment to Company "B" and reassigned to Company "A." Lt Stoudt was assigned to Company "B" and given command of the 2d Platoon. Pvt Huntington was discharged [from] the Service to accept commission as 2d Lt, CE, and assigned to 150th Combat Engineer Battalion.[7]

WEDNESDAY — 27 DEC 44

. . . .

Personnel — Sgt Rohrmann was discharged [from] the Service to accept commission as 2d Lt, CE, and assigned to 145th Combat Engineer Battalion.

The news continues to improve. The garrison at Bastogne was relieved last night, after holding out against superior German forces heroically. The Germans have been stopped about four miles short of the River Meuse and on the northern flank of their thrust they are digging in south of Liege. Third Army continues fighting forward on a 20-mile front. American planes had a field day today, for the fifth day running.

THURSDAY — 28 DEC 44

. . . .

No damage to road net by enemy aerial attack past 24 hrs.

The Germans appear to have become defensive rather than offensive, and reports have it that General von Rundstedt is attempting to regroup and reform his forces for another thrust.

FRIDAY — 29 DEC 44

. . . .

Road maintenance — Deicing resumed; stockpiles being established on MSRs. N7, Luxembourg City to Ettelbrück, reconnoitered and found in good condition. No preparation has been made to build the bridge across Alzette River. No damage to road net by enemy action past 24 hrs.

It is interesting to note the Message sent to S-3 by the Battalion Commander on the defensive setup for the 1st Battalion which was the purpose of our coming up when the Germans ruined the First Army. "This Battalion has no FPL" [final protective line?].[8] This is because the sector is ten miles long and insufficient men and weapons are available to cover and mutually support this line. Rather, the defense decided upon was to place road obstacles at strategic points in the entire road network. These obstacles are covered by fire. However, they are too far apart to mutually support each other, the support coming only by signal communications." The Germans continue raising hell in their bulge.

Personnel — Tragedy struck the Company today. Enroute to the 1st Battalion [Medical] Aid Station at Differdange with the sick call, Tec 5 Looper's 2–1/2T truck skidded on the railroad overpass on Ic26 about one mile SE Differdange, fell to the railroad tracks about 20 feet below and turned over, killing Zemla, injuring James Jones, seriously injuring Abels, Looper, Willie Barr (attached colored driver), and "Butch," the little French boy who has been a mascot to the Company since we were in Bar-le-Duc. The French boy is not expected to live.[9] All soldiers were dropped from strength upon hospitalization.

SATURDAY — 30 DEC 44

. . . .

No damage to road net by enemy attack past 24 hours. The site of yesterday's accident has been marked with warning signs.

Personnel — Pvt Hughes was reassigned to and rejoined [the] Company. The investigation declared the injury was in the line of duty.

Although the danger from the German Bulge is by no means completely over, the counterattacking forces — especially the Third Army — have contained the offensive and it is safe for the strategists to speculate and generalize. General von Rundstedt is said to have miscalculated when he failed accurately to estimate our capability to move. General Patton moved his Army miraculously. For instance, the 80th Division was leaving the line for a rest when orders were changed to defend the City of Luxembourg on the north. They traveled the 150 miles and were dug in within 15 hours, a feat of logistics if ever there was one. Meanwhile, the Russians have surrounded Budapest and are holding a good part of the city.

SUNDAY — 31 DEC 44

. . . .

Establishment of the 1st Battalion bridge material dump at Colmar is 35% complete: materials are being moved from Command Post to the dump which is being guarded by the nearby bridge guard of the 166th Engineer Combat Battalion.

German aircraft bombed road junction N3-N13 about 1000 hrs; reported by witness, Lt Stoudt.

Personnel — Sgt Parker was transferred to Battalion Headquarters as the Battalion Construction Foreman which means a boost for him to Staff Sergeant's rating.

Third Army is attacking north of Bastogne, successfully sealing off the tip of the German salient.

MONDAY — 1 JAN 45

. . . .

Deicing operations continue: stockpiling N13 (Bettembourg–Frisange), N63 (Volmerange–Altzingen), and the road Dudelange–Kayl. No damage to road net by enemy action past 24 hrs.

Personnel — Pfc Edmunds was reduced to grade of Pvt; the following soldiers were promoted to Private First Class: Bullock, Garcia, Merritt, Miller, Minkemann, Peterson, Thornton Williams, and Isham and James Wilson.

Third Army has taken the town of Rochefort and continues to attack on a 20-mile front between St Hubert and Bastogne. The Germans are on the defensive in the territory they took in their offensive which began 17 Dec.

TUESDAY — 2 JAN 45

Bridge guard — Relieved of Luxembourg City bridge; assumed responsibility for two bridges in Esch, one in Hayange [Harlange?], and one in Tiercelet [?].

Deicing operations continue unchanged.

Completed hauling all materials to 1st Battalion dump, Colmar. Battalion

Headquarters and "A" and "C" Companies moved to the chateau at Ansemburg.

No damage done to road net by enemy action past 24 hrs.

WEDNESDAY — 3 JAN 45

. . . .

Personnel — Promoted to Sergeant were Boling and Peterson and Shouse; to Corporal Chavez, DeMattei, and Jaspersohn; to Tec 5 Hetler, Lemp, and Whitaker.

Third Army is five miles past Bastogne, and gaining daily. The Germans are replacing their Panzers with the more-economically-sacrificable [*sic*] infantry divisions. Some 20,000 prisoners of war have been taken in the past two weeks, and 450 tanks and self-propelled guns have been knocked out.[10]

. . . .

FRIDAY — 5 JAN 45

Bridge guard — Because we have not been properly relieved of responsibility, we did not turn over the bridge guard posts to Battery B, 734th Field Artillery Battalion who came out to relieve us.

Deicing operations resumed full scale because of the snow and slush. Worked on B4 and Gc63 and routes through Esch. Cleared snow and slush [*sic*] off roads. Road net not damaged by enemy action past 24 hours.

Personnel — Cederberg was dropped from rolls upon being hospitalized; Koontz replaced Wiszowaty on Temporary Duty at Regimental Headquarters as guard; Fisher was arrested in town for drunkenness and disorder and is in arrest pending Court-martial.

There is not much change in the news. The Third and First Armies are successfully attacking the ten German divisions in the Bastogne corridor.

SATURDAY — 6 JAN 45

Bridge guard — Relieved by the 734th FA on bridge guard, we nonetheless continue maintaining a darkness guard on the two bridges in Esch.

Deicing continues. Sanded N4, N13, Gc63, and road Dudelange to Esch.

Road Gc63 (Bettembourg to Volmerange) swept for mines and posted. No damage done to road net by enemy action past 24 hrs.

Personnel — Pvt Fisher restored to duty from arrest, still pending CM [court-martial].

SUNDAY — 7 JAN 45

Darkness guard, Esch, continues unchanged.

. . . .

Personnel — The six carpenters returned to duty today from Temporary Duty with 1134th Engineer Combat Group at Esch. Sgt Hawke was replaced as Motor Sergeant by Sgt Bennett, and Hawke took over the other job as Foreman Mechanic.[11]

The Colonel ordered the 1st Battalion at 1500 hours to prepare for demolition [work on] the roads east of Battalion Headquarters — N7 north of Luxembourg City to Mersch and the supporting roads — in the event it becomes necessary to blow bridges and culverts should the enemy counterattack the two-division thrust being launched in front of us. There is a gap between the divisions for the beginning of the attack, and after what happened to the First Army for not being prepared, everyone is scared of another spanking . . . from the enemy and from SHAEF. "A" and "C" Company have been assigned this job.

The arctic-like front continues to show [*sic*, slow] American advances[.] The First is fighting with Britain's Montgomery, the Third continues strong in the south. The First have cut one of the two German supply arteries, and only 12 miles separate the two armies.

MONDAY — 8 JAN 45

Bridge guard — Relieved of darkness security guard on the two bridges in Esch at 1800 hrs.

Deicing consists of snow removal with grader and sanding N4, N13, Gc63. No feeder roads were worked on today.

Replaced man-hole cover on N7 in Luxembourg City. No damage done to road net by enemy action past 24 hrs.

The barrier zone was hot, then cold, in less than a day. Last night a message dated 2345 arrived for the Battalion Commander. "Demolition will be executed on Corps order or on mass movement of enemy vehicles. Keep your men out until further orders from this headquarters. We expect to send a messenger to your hq some time after 0200 with further information. If demolitions are blown they must be defended as long as possible. It is no barrier if enemy is free to repair it." Only Major Polick, the S-3,[12] could be capable of so gilding the lily.

In the morning about 0600 the Battalion Commander ordered "A" and "C" Companies to man the defenses with a minimum of manpower, keeping the balance alerted. Deicing was to be continued. That's the way it is: do two jobs with one supply of manpower and with no indication as to the relative importance of the jobs.

Finally, about noon, the orders came through to remove the explosives and stockpile them in a location readily available for instant use. So "A" and "C" Companies, after spending the night of the 7th preparing their positions, removed them.

Came 1730 and the payoff: replace demolitions immediately. Another night went to the placing of charges. At 0505 the harried Battalion Adjutant announced completion of the mission.

Snow fell all day: roads are icy, visibility poor.

TUESDAY — 9 JAN 45

Deicing continues. Grader and 4T plow removing snow from N13 and Gc63 from RJ N13 to Ottange (new MSR) RJ N4-N13 sanded for 100 yds in all directions. N4 is in good condition. The supt [superintendent] in the steel mills requests we stop hauling away granulated slag from present location of shovel as it is needed elsewhere. Lt Stoudt made arrangements to draw 1,000 T of slag at the rate of 40–50 T daily: this is plant capacity. Source is the Hadir Steel Works, Differdange. Clamshell is needed to load the trucks from out of railroad cars.

At 0900 we relieved Company "F" of their barrier zone. All demolitions except those on the bridge removed and stockpiled and are guarded by an NCO and 11 EM [enlisted men] with transportation. The bridge is being guarded by the same crew.

No damage to road net by enemy air attack past 24 hrs.

The Americans landed on Luzon today, the principal island of the Philippines, from a task force of 800 ships and 2500 landing craft of all kinds. Penetrations were made on a 15 mile front with minor casualties.[13]

The Germans continue to retreat as troops from the south and north battle in the arctic weather, without air support, to close the gap between the First and Third Armies.

WEDNESDAY — 10 JAN 45

Reconnoitered the barrier zone which we took over last night and enlarged the plan to include three craters as indicated on report overlay. Requested permission to dig the holes and then, if they are not used, to refill them when the danger is past. Other work of placing obstacles is in progress.

The demolition plan calls for cutting in at least one place all main arterial roads leading into Luxembourg City and the feeder roads of these arteries using mine fields and dummy mines or craters. Mines are preferred, but supply is not known. Have requisitioned 40 in addition to the 30 issued as authorized. Advice must be forthcoming before morning in order to execute planned work.

Reconnoitering area for a new Command Post vicinity Kersch. When we wanted to move there three days ago, permission was denied.

No damage to road net by enemy air attack past 24 hrs.

THURSDAY — 11 JAN 45

No deicing work done today.

The new defense plan which we have devised is being worked on but is not complete. Meanwhile, that taken over from Company "F" is in readiness should one be needed before our improved one is completed.

The 2d Platoon moved forward and opened new billets in the Chateau at Mersch where the Company will move tomorrow.

No damage to road by enemy air attack past 24 hrs.

BIVOUAC 21 — MERSCH, UP8228, NORD DU GUERRE ZONE.

ARRIVED 1230 12 JAN 45 — DEPARTED 1030 12 MAR 45.

FRIDAY — 12 JAN 45

The first Platoon was left behind to complete the defense plan while the rest of the Company moved to Mersch. Completion of barrier zone estimated for 1800 tomorrow.

No damage to roads by enemy air attack past 24 hrs.

Personnel — The 1st Sgt, Douds, was awarded the Bronze Star Medal for "meritorious service" at Regimental Headquarters today.[14]

SATURDAY — 13 JAN 45

The 1st Platoon arrived at 1230, and now the entire Company is present in Mersch.[15]

Working conditions and unavailability of earth auger until late prevented completion of barrier zone as estimated. Instead, work will be done before 1800 hours tomorrow. The plan is corrected: positions 3 and 10 deleted; 2 moved west; 9, 10, and 11 added; further corrections — 2 contains 15 AT [antitank] mines instead of 11 and 4 contains 25 AT mines instead of 11.

No damage to roads by enemy air attack past 24 hrs.

SUNDAY — 14 JAN 45

The barrier zone was completed by 1800 hrs. All demolition and minefields can be effected within four hours.

Deicing MSRs as required: worked N7, N8, N11, and N1. No damage to roads by enemy air attack past 24 hrs.

Personnel — Platter hospitalized, LD undetermined. The following new men joined the Company: Manske, Murray, Rajewski, Gabler, Sharpe, Begale, and Bordonaro.

The Germans have been pushed back into Germany.

MONDAY — 16 [SIC 15] JAN 45

Improving the barrier zone. All explosives stockpiled at CP. The grid coordinates for obstacles B7 and B11 as reported in the Consolidated Regimental Defense Plan 13 Jan are wrong. Correct coordinates were given [by] the S-3.

Deicing — N7 is clear of ice. Feeder roads and other routes deiced as required. The VO road Plankenhof to Fishbeck sanded to facilitate the movement of tanks.[16]

Test-fired sample of each bundle of explosives in the charges for barrier zone. All explosives satisfactory.

No damage to roads by enemy air attack past 24 hours.

TUESDAY — 17 [SIC 16] JAN 45

Continue improving barrier zone. In the AT mine fields only the dummy mines are actually in place.

Continue patrolling and sanding roads as required. N7 is clear of ice. Other roads in good condition. No damage done to roads by enemy air attack past 24 hours.

At a formation of the entire Company, Capt Manuel F Gonzalez relinquished command of Company B to 1st Lt Earl E Hall. The Captain is relieved and assigned as Commanding Officer, 1st Battalion, replacing Major Fred G Mylius, relieved for reasons of health; he departs in the morning.[17]

WEDNESDAY — 17 JAN 45

Continue improving barrier zone. One abatis and one crater have been added. All obstacles and minefields are being posted with symbols and unit designations. This is to enable Army inspectors riding past to identify defensive works. This, without the shadow of a doubt, is essential war work.

Deicing continues. In addition to the old areas, we are now responsible

Capts. Manuel F. Gonzalez, George R. Sant, and Robert E. Bolle at Eschnerich. All three men, company commanders in the 1st Battalion, 1303rd EGSR, were general foremen or owners of construction companies in civilian life, had considerable experience in the business, and went through OCS together. They were close friends. Courtesy Dwight Gowdey

for a forward area, work on which started 1300 hrs. No enemy aerial damage to roads past 24 hours.

Wrote Major Polick to the Battalion Commander: "The answer to the barrier zone is that the thing will gradually die a natural death. There is no change at present but as things go on we will gradually decrease our barrier zones until they are all eliminated."

THURSDAY — 18 JAN 45

Barrier zone improvements and marking 100% complete.

Deicing materials paced on [highways] . . . in forward area. A snow fence established on N26. Anti-skid material also paced on 5th Division MSR between N1 and N14. No damage to roads by enemy air attack past 24 hrs.

Personnel — Pfcs Bates and Silvers exchanged with their respective command cars between Company "B" and Headquarters 1st Battalion as drivers for the commanders concerned, because Captain Gonzalez wanted Bates.

Our new roads are to be kept open for the XII Corps attack which is jumping off tomorrow morning. The Germans have been pushed back into the Siegfried Line. The Russians have taken Warsaw and are within 12 miles of the German border.

Capt Bolzer, Adjutant, 1st Battalion, was transferred out of the Regiment. The new Adjutant is 1st Lt Bubb, of Company "A."

FRIDAY — 19 JAN 45

All bridges and large culverts on N7 — Luxembourg City to Ettelbrick — and on N26 — Mersch to Colmar — inspected, classified, and posted. Two under classified bridges were re-classified 1w70–2w40.

All roads in the forward area were checked during the night and continuously during the day. Deicing unnecessary because of the weather: it is foggy and rainy and slushy and snowy.

We are hauling 20 I beams from Differdange to the bridge site at Ettelbrück for Company "A." Diekirch taken by Americans today: we will build the bridge there.

No damage to roads by enemy air attack past 24 hours.

SATURDAY — 20 JAN 45

Deicing patrols operated all day. Sanded N7, Mersch to Luxembourg City. Responsibility for forward area ceased 1800 hours tonight; the attack has been launched successfully.

Bridging — a. Kopstal bridge construction begun today. Cleared site, hauling materials. All decking. 12×12 timbers, and most of the 6×6 timbers are at the site. Steel will be hauled from Esch.

b. Mersch bridge approaches being repaired on N26.

c. Of the 20 I beams we are hauling for Company "A" bridge at Ettelbrick, 14 were delivered to their site by 1800 and the remaining six left at the CP.

d. Delivered Bailey parts and tools to "C" Company at Ettelbrick. They will remove the Bailey instead of us as was planned before we got the bridge building jobs.

e. The second bridge assignment is at Diekirch. Orders received too late to make daylight reconnaissance today: will make it in the morning. The infantry is still clearing the town of snipers. Booby traps are everywhere, and the bridge site is extensively marked.

No damage to roads by enemy air attack past 24 hrs.

Finally, that "natural death" of the barrier zone became a reality. Messaged the Battalion Commander: "Men and explosives may be removed from barriers in the defense plan to such an extent that they can be replaced and manned on 8 hours notice."

SUNDAY — 21 JAN 45

Bridging — . . . .

d. Diekirch bridge reconnaissance made, and the work party — Lt Stoudt's 2d Platoon — moved [to] the site today. Work starts in the morning. Timbers are being hauled from Esch. A steel detail was sent to Esch also, but the stock we want is buried beneath 100 other beams which have to be moved. Two, possibly three trailers will be loaded some time tonight, of which two are expected at the Command Post around midnight. The heavy equipment we borrowed from "A" Company cannot be returned before noon tomorrow because of this delay and a delay occasioned by Army Engineer in clearing the steel for us.

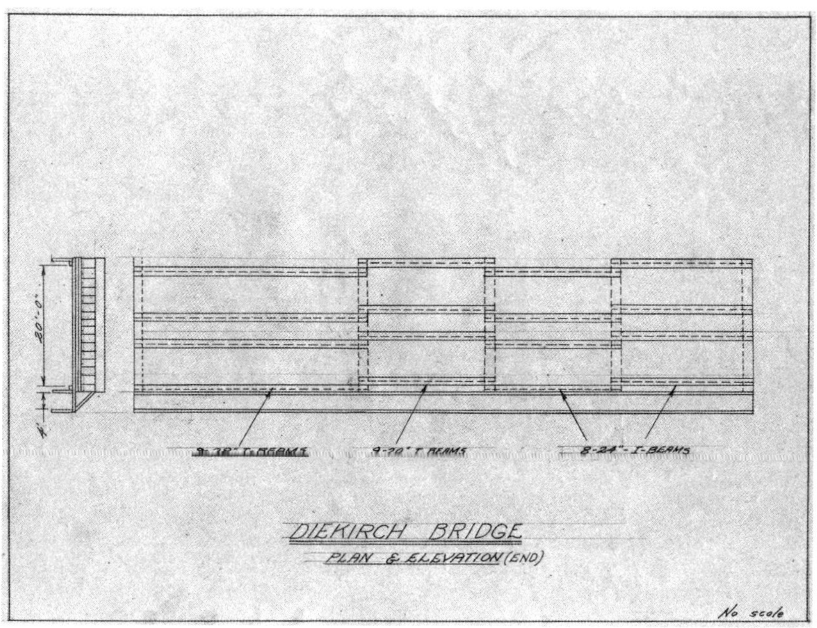

Drawing of stringer-beam placement for Diekirch bridge. Courtesy Earl E. Hall

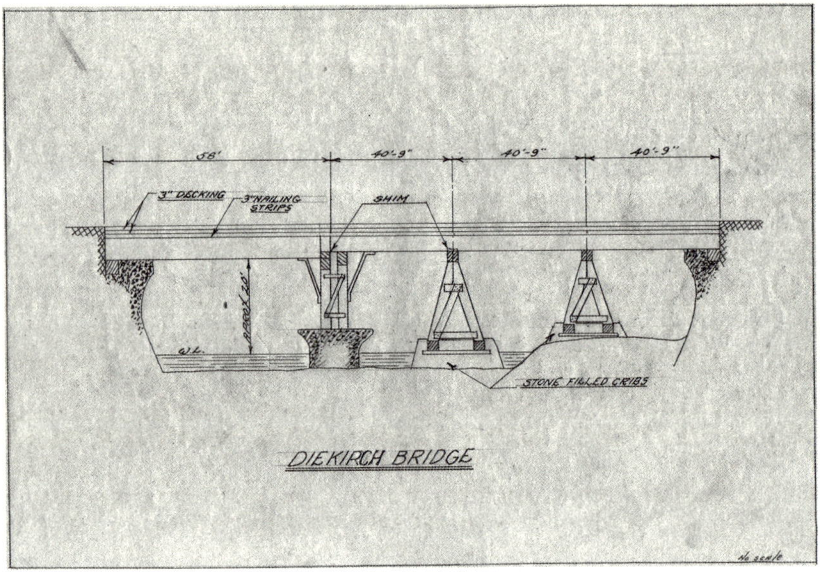

Drawing of Diekirch bridge. Courtesy Earl E. Hall

MONDAY—22 JAN 45

Road work continues: graded and sanded N8, N26, and other military roads. The undelivered beams remain undelivered: no transportation.

Bridging—. . . .

b. Diekirch bridge—The abutments were found mined: two rows of 11, each, Tellermines. The beach below the abutments contained four rows of Schu mines (25) which had to be removed. Partially finished with the hand removal of these latter mines, Lafferty stepped on one and blew off his left foot. Also wounded about the face were Dziwenka and Kaczanowski, and Mize, H&S [Headquarters and Service] Company dozer operator attached, was slightly injured about his face. All men were hospitalized, dropped from the rolls. Lafferty and Dziwenka, insisting on outlining the pattern of the field to their squad leader, Sgt Pennington, who rushed to their aid oblivious of his own danger, demonstrate a courage and fortitude unseen before as they refused to be evacuated until the other men had the pattern in mind to finish the work. The Company has recommended these three for Silver Stars.

COMPANY B STARTED WORK on the Diekirch bridge on 22 January, with clearing of Schu and Teller mines from the banks and in the streambed where the piers would go. The task was dangerous, with the above-named four men injured in the task. Only one existing stone pier was usable; it was leveled off and a timber bent was placed on top of it. The engineers constructed two piers using timber cribs on top of the debris at the bottom of the river. "A steel frame gin-pole, left by the Germans, was utilized to place the steel beams." Warming weather brought a rapid rise in the water shortly after construction started, but the crib piers were already in place. Nevertheless, they had to be built higher with added stone to keep them from washing out. Work continued despite the flood. The floating bridge at Diekirch was closed, "the treadways, floating bridges, and Baileys in the vicinity were either washed out or inundated. This limited very considerably the number of crossings of the Sauer which could be used and made it essential that this high level, fixed bridge be opened at the earliest possible time. Accordingly the job was pushed on a 24 hour schedule and the bridge was opened for traffic . . . two days ahead of the scheduled completion date. From the minute it was opened there was a continual stream of traffic resulting from the changing troop dispositions which had awaited the completion of this bridge." The area was shelled numerous times during construction, but the work was not suspended. No casualties resulted from the shelling, though 2nd Lieutenant Stoudt and two men were injured while placing steel beams.[18]

TUESDAY — 23 *JAN* 45

Road work continues on N8 and the VO from Mersch to Kopstal: graded and sanded these roads. No damage by enemy air attack to roads past 48 hrs.

Bridging — a. Kopstal bridge completed 100% today, as estimated. This includes reinforcing the adjacent bridge. The traffic island will be built tomorrow.

b. Diekirch bridge — Additional mines were found and had to be blown in place for clearance. Half-pound blocks of TNT 18 inches apart having been found inadequate, the Tellermines were used to demolish the minefield in place. Enemy artillery fired on the site during both the day and night. Work is continuing, nonetheless. Both abutments are being formed, and one bent is being constructed. The last six I beams and 700 pieces of decking are at the site.

The Third and First Armies met today. The Third has taken St Vith. The Russians are within 15 miles of Breslau, within 20 miles of Konigsberg in East Prussia

WEDNESDAY — 24 *JAN* 45

Graded and sanded N8 and the VO road, Mersch to Kopstal. No damage to roads by enemy air attack past 24 hours.

Kopstal bridge traffic island continues.

Diekirch bridge abutments are prepared, one bent is done, and the cribs are being built. Steel from an adjacent unfinished bridge is being salvaged and welding of the steel members for pier is in progress. Estimate completion for 5 Feb: the job is 15% done as of today. Enemy artillery shelled the sight [*sic*, site] last night again: 15 rounds of probably 88 mm fell in.

Personnel—Tec 5 McGlashan hospitalized with pneumonia and dropped from the rolls.

THURSDAY—25 *JAN* 45

Graded and sanded N12 and the Mersch–Kopstal bridge. No damage to road nets by enemy aerial attack past 24 hours.

Kopstal bridge traffic island completed 100% today. Diekirch bridge continues abuilding. One abutment sill seated, one crib completed, a pier bent in place but not erected. Completed welding bracing members and sent them out to the pier. Job 20% done. The shelling has stopped.

The weather is bad again: snowing. The Russians have reached the Baltic, cutting off the Germans there. They are fighting in Breslau. The territory gained by the December offensive has been cleared entirely of Germans.

FRIDAY—26 *JAN* 45

Added to the roads worked on yesterday N8. No enemy damage to roads from the air during the past 24 hours. Late last night our R-4 cleared three miles of road for the 5th Division north of Diekirch.

Today we hauled 32 12" I beams for Company "D" bridge starting at Echternack. The Ninth Army jumped off, heading for the Roer River.

Diekirch bridge progressing: the first crib is being rock-filled, the second one is being built, one bent is erected and the second one begun. The job is 24% done.

A 5th Division officer found some explosive coal at Diekirch near the bridge site: pressed coal containing firing devices inside, beneath the trademark.

Personnel—Tec 5 Lemp finally got his request answered favorably and has gone home to his old outfit, the 326th Airborne Engineer Battalion.

SATURDAY—27 *JAN* 45

. . . .

Personnel—Sgt Pennington sent to Paris on a three-day pass.

Third Army is on the Luxembourg-German border [*sic*].

SUNDAY—28 *JAN* 45

No change in road maintenance, and no enemy damage. Diekirch bridge. . . .[19]

Both cribs are completed to sill height. One crib is filled and the sill for the bent seated. All A frames for tower bents finished, also finished the second bent for the double pony-bent. Job 30% done.

Personnel — Reed. 1st Battalion Headquarters Detachment, transferred to the Company, and Lange was reassigned to and rejoined the Company.

The Russians have taken Memel, are shelling Breslau, Königsberg, and Poznon. The weather continues bad: snow flurries and treacherous icy roads.

MONDAY — 29 *JAN* 45

. . . .

However urgent his request, we cannot send G-2, Army, a sample of the previously reported explosive coal because the 5th Division officer only showed it to us and there are no further specimens to be found in the coal pile.

Personnel — Manke hospitalized and dropped from rolls. Major Carr, former Commander 2d Battalion, visited the bridge site today. Third Army is a mile from the Siegfried line and going strong.

TUESDAY — 30 *JAN* 45

Continue grading snow and sanding ice on N8, N12, and N22. No enemy aerial damage to roads past 24 hrs.

Diekirch bridge — Set the sill for second crib and partial: erected the tower bent on it. First span steel completely set. Cross-bracing the steel in place. Nailing strips are placed. Set the second pony bent on rock crib and bound the two bents together. Job 40% done.

Completed road reconnaissance reported to Battalion Hq today.

The Eighth US Army has landed on Luzon without casualties. Where Germany, Luxembourg, and Belgium meet on the Our River, Third Army has a four-mile front that is raging.

WEDNESDAY — 31 *JAN* 45

No change on roads, either in damage or in maintenance.

Diekirch bridge — Decked first span. Completed second tower bent. Setting steel in fourth span. Job 45% done.

Personnel — Sgt Peterson was awarded the Soldier's Medal for heroism in Dudelange when we were there and he helped fight a gasoline fire one night. The presentation ceremony took place at Regimental Headquarters. S Sgt Sheridan, who pinched his foot between two I beams at Diekirch, was hospitalized against his will and dropped from strength. And Schara was assigned and joined the Company today.

THURSDAY — 1 FEB 45

Diekirch bridge — Completed second course of decking on first span, set five I beams in fourth span, one I beam in second span, and added an additional three feet to the height of the rock-filled cribs. The job is 50% complete.

. . . .

All enlisted men younger than 31 years were given physical exams to determine their fitness for retraining as infantrymen: all units are losing men this way to fill up infantry regiments whose losses have been great.[20]

The weather is filthy. It is turning warmer and a general thaw has been felt on both the Eastern and Western fronts. Many places are reported quagmires. Rain has been falling and melting the snow; rivers are up and threatening bridges. The critical bridges are being carefully watched.

Third Army advances across the Our River on a seven mile front three miles deep.

Personnel — S Sgt Sheridan rejoined the Company today.

FRIDAY 2 FEB 45

On the Diekirch bridge the fourth and second spans have all steel set and braced, and the first course of decking is on the second span. Bridge 70% complete.

Always a "hot" job, the Diekirch bridge became very hot suddenly. The S-3 messaged the Battalion Commander: "Corps reports that the Sauer River is rising fast and their floating bridges are endangered. Would like you to rush the Diekirch bridge a little harder and beat your estimated completion date of Sunday morning." Capt Gonzalez wrote him: "There is no way to push Diekirch 'a little harder' as we are already doing all possible and have been. Estimated completion date is not Sunday, but 5 Feb 45. I did remark that with luck we might have traffic over the bridge Sunday. We will do our best and open the bridge as soon as possible." But to Lt Hall he said: "Diekirch must be open tomorrow as early as humanly possible. Don't spare the horses! Suggest you carry on through night even with steel, if crews finish decking before day. Roads require no immediate attention so you should have plenty of manpower. The 80th and 5th plan to use the bridge."

All bridge[s] in Company area reconnoitered, in good condition. The Mersch bridge, though, will be endangered if the water rises another foot — it will be hitting the steel.

In the midst of this rat race came the word to send one man on a three-day pass to Paris: each Company send one man, the Battalion one officer.

Sgt Pennington was chosen, for his work in the minefield several days previously.

Personnel — S Sgt Dunbar, Tec 5 Ray, Pvts Begale and Gruppi were all hospitalized and dropped from the rolls. Tec 5 McGlashan was reassigned to and rejoined the Company.

SATURDAY — 3 *FEB* 45

The Diekirch bridge was opened to traffic at 1930, 95% completed. The original completion estimate changed yesterday from the 5th to the 4th. Classification unchanged. The bridge lacks the second course of decking on the third span, handrails, and curbing.

This morning Lt Stoudt was seriously injured when he was struck by an I beam suspended between two cranes and being placed.[21] A fractured pelvis and internal bleeding prevent a diagnosis for 24 hours. Staff Sergeant Cooke carried on without an officer again, and successfully completed the bridge ahead of schedule.

New crossings of the River Our by Third Army reported. The First has advanced three miles on a 30 mile front into the first wave of Siegfried Line defenses. Yesterday 1,000 Forts plastered Berlin with 2500 pound bombs at midday.[22]

Personnel — Lt Stoudt dropped from rolls.[23]

SUNDAY — 4 *FEB* 45

The Diekirch bridge is 100% complete. Additional cribbing is planned and will be placed when water permits.

Road maintenance resumed: filled chug holes with gravel within Company network of responsibility. Also maintaining Timber Trestle bridge at Wolferdange (N7, between Mersch–Luxembourg City). The "Santa Fe" Doughnut-Clubmobile serviced the Battalion today. Announcement was made today that after two years Gen MacArthur reoccupied Manila freeing some 2,000 prisoners.

The First and Third Armies are fighting in the Siegfried Line, the First having completely penetrated the first barrier. Colmar is occupied by the Allies who hold 50-mile front on Rhine.

MONDAY — 5 *FEB* 45

The walkway on the Diekirch bridge is being redecked. Additional work on cribs is in progress. And road and bridge maintenance continues.

This morning Lt Gen Patton personally pinned the Bronze Star Medal on Lt Stoudt in the hospital on a tour of inspection, first thoroughly "chewing" his aide for not having available any medals in his pocket and making the harried Colonel return to Army Hq in Luxembourg City to obtain same.

Lieutenant Moseley was promoted today to full lieutenancy, effective 1st February.

TUESDAY — 6 FEB 42

Completed the Diekirch bridge walkway and the additional cribbing.

Maintaining Bailey Bridge and fixed bridge approaches within area of responsibility. Continue road maintenance.

. . . The Colonel messaged [the Battalion Commander] that after a transit check the Diekirch-end pier had settled and, though we [*sic*]was not yet worried, it needed watching. A second reading was promised for tomorrow, and direct notification if there was further settlement. The Captain wrote back: "The difference in elevation between the first pier from north end of Diekirch bridge and the other piers was known before the bridge was complete, but after the span was decked. The settlement occurred after the river rose but before traffic was over bridge. A check after opening of bridge revealed no further settlement under traffic. My opinion is that settlement was caused by consolidation and compaction of crib core caused by soaking action of water — not by scour under crib and will not get worse and is not dangerous. Lt Hall wanted to correct with shims prior to opening bridge. Since, in my opinion, this difference in elevation affects nothing but appearance of bridge, and since the work involved correcting was considerable, I instructed Lt Hall to leave it alone and only to watch it to see if settlement increased. Incidentally, Lt Hall is carrying his handrail across at a uniform elevation (not following sag) which should improve the appearance."

Following all of which he wrote to the Company Commander: "Don't forget to paint wheel guards on bridge at Diekirch."

WEDNESDAY — 7 FEB 45

Completed filling additional cribs at Diekirch bridge. Continue bridge maintenance and road work. Hauled materials for the repair of Lintgen and Christnach bridges.

Personnel — Ray was reassigned to and rejoined the Company.

The bridge completed by Company "C" near Diekirch collapsed following the faulting of one of the masonry piers.

Reconnoitered site for 65th Field Hospital: insane asylum of great ugliness, partly completed, in deplorable condition now. Need window replacements, carpentry work to partition large rooms, and cleanup of rubbish. Work starts in the morning.

Co "D" had five men injured by artillery at Echternach. Third Army jumped off again this morning: two divisions of XII Corps and two of the III [*sic*]. Ten crossings were made. The going was tough and two platoons of one combat battalion of Engineers (believed to be the 150th) were lost.

THURSDAY — 8 FEB 45

Hauled 50 assault boats and 900 paddles from AESP 13 to Corps Dump at Junglinster on N11. The last 100 paddles were unavailable. Hauling cement from Toul to Battalion Headquarters.

Continue road and bridge maintenance, and demudding N7. Started readying the site for 65th Field Hospital in Ettelbück. We will not have to do the cleanup, but only Engineer work.

Bridging — Repairs of Christnach bridge started: the Lintgen bridge will be started after all materials are hauled.

The Regiment was awarded campaign stars for participation in Normandy and Germany [actually Rhineland] campaigns. The campaign Northern France is not yet announced.

FRIDAY — 9 FEB 45

Continue road and bridge maintenance. The Christnach bridge is 100% repaired. . . .[24]

The Lintgen bridge is 75% repaired.

The hospital site is progressing: building partitions, installing windows, and starting to wire.

The cement arrived from Toul at 1000 and was sent on to Battalion dump.

Third Army is within a half-mile of Prum. The Our River crossing between Luxembourg and Germany continues to be "difficult."

Personnel — Stagner was hospitalized and dropped from rolls.

SATURDAY — 10 FEB 45

Lintgen bridge repairs competed 100%. . . .[25]

Hospital job 70% done. Road and bridge maintenance continues.

The use of Luxembourg civilians to do road work is planned.[26] The railroad will deliver carloads of crushed rock to an approved siding near Mersch, and our job will consist of supervising the unloading and stockpiling of rock here, and its eventual use on the roads. The bad spot on N7 one mile south Ettelbrück is being cut out and refilled with heavy rock ballast.

The Red Cross doughnut wagon was here today again.[27]

SUNDAY — 11 FEB 45

Continue road and bridge maintenance, concentrating on the bad spot on N7.

The hospital is 75% complete. Windows installed and blackouts rigged. Additional work on quarters of personnel to be done at request of Hospital Commander.

The entire Company saw the required film INSIDE GERMANY this morning.

The weather continues filthy. "General Mud" rules supreme.

MONDAY — 12 FEB 45

The hospital in Ettelbrück is 85% done. Bridge maintenance continues unchanged. And road work, in addition to the quagmire now includes giving Class A attention to the following first priority roads: N7 Mersch to Ettelbrück (including the stretch of N26), thence to Diekirch. N16 Diekirch to left border. N27-N7 to left border. N14 from road junction N14-N9 to Diekirch. N9 Mersch to Fels.

The use of civilians for road work has been approved. Lintgen is the siding for the arrival of gravel and rock. Starting tomorrow morning trucks will pick up 200 workers for each of the two battalions at the railroad station in Lux City at 0745. They come from Esch. The train departs Lux City at 1730 daily. No formal attendance records will be kept, but a daily report will be submitted listing the number of men who worked. Timekeepers are civilians, too. The noonday meal will be a C ration.

Third Army has Prum, the German highway center, and is improving the Sure and Our River crossings. The Crimean Conference — Roosevelt, Churchill, Stalin — ends today.

TUESDAY — 13 FEB 45

The hospital is 90% done, but we started installing plumbing facilities, too. Road maintenance continues as yesterday (used 36 civilians today), and bridge maintenance is unchanged. "The road leading from Diekirch east (N17 & 19)," wrote the Battalion Commander, "to your forward boundary is playing a key role in the progress of the front. Will you jump on that baby with both feet and put it in shape. If work is needed farther forward and you can take care of it, hit that, too."

Third Army has completely cleared Prum.

Personnel — Pvt Chalom hospitalized, dropped from rolls.

WEDNESDAY — 14 FEB 45

The hospital is 92% done. Bridge maintenance continues unchanged. Road maintenance continues unchanged, using 50 civilians. All Bailey Bridge parts at Gilsdorf, Bettendorf, Moesdorf, and Reisdorf are being collected and hauled to Company "D" at Echternach.

Clear weather today permitted extensive use of the air arm: more than 5,000 sorties were flown. This is perhaps the heaviest concentration of air-power since D-Day. Dresden and Chemnitz received the heaviest poundings. Third Army extended its Echternach bridgehead one mile on a six mile front.[28] The 2d Battalion started building the Echternach bridge.

THURSDAY — 15 FEB 45

In addition to previously reported work, today gravelling of driveways and walks was begun for the 65th Field Hospital and the job is 95% done. No change in road and bridge work.

Yesterday we hauled to Company "D" the following Bailey Bridge parts: 46 panels, 96 panel pins, 14 sway braces, 16 rakers, 35 chess, 22 ribands, 5 bracing frames, 12 transoms, 32 bracing bolts, and 28 riband bolts.

56 civilians worked today.

It was reported today that 1200 carrier-based aircraft attacked Tokyo today in one of the largest task forces ever assembled.

FRIDAY — 16 FEB 45

Completion date for hospital job moved up one day. Gravelling and plumbing continue: 98% done.

Continue maintenance of all roads and bridges assigned. Five cars of gravel received at Lintgen. Worked 57 civilians.

The following bill of materials for an "average" bridge was submitted to Battalion Headquarters as basis for planning a Battalion Bridge Dump: 32 20" I beams, 55 12×12×14' timbers, 432 6×6 curbing, 875 3×3×12' decking, 600 3×3×12' cribbing; 810 3×3 railing, 12 kegs 60d nails, 2 kegs 80d nails, 1 box (30) 24" bolts, 6 boxes drift pins, 2 gals white paint, 2 4" paint brushes.

The carrier-borne aircraft continue attacking Japan for the second day running.

Personnel — S Sgt Dunbar and Pvt Gruppi reassigned to and rejoined Company. Starting to carry limited assignment personnel on the M/R [morning report], we have 11 enlisted men in this category.

SATURDAY — 17 FEB 45

The 65th Field Hospital site is 100% complete and ready for occupancy.

Continue maintenance of all assigned roads and bridges. Worked 57 civilians today.

The Red Cross Clubmobile was here today, not only at the Command Posts but to all the jobs, too.

Personnel — Pvt Wiszowaty was hospitalized today and dropped from rolls. His return this time is not contemplated.

SUNDAY — 18 FEB 45

Additional plumbing is being installed at the 65th Hospital at the request of the Commander.

No civilians showed up for work today, it being Sunday. We continued road and bridge maintenance. No gravel arrived. Started hauling 1000 smoke pots out of the mud near the river at Mersch for the 91st Chemical Battalion.

The attacking Third Army gained a mile and a half on a 28 mile front and considerably improved some positions. The First and Ninth Armies are waiting for the Roer River to subside before crossing in attack: the flood were [*sic*] caused when the Germans opened the dams on the headwaters of the Roer.

MONDAY — 19 FEB 45

Continuous road and bridge maintenance. Eight cars of gravel delivered, of which six and one-half have been un-loaded. 61 civilians worked today.

Continue hauling the smoke pots out of the mud. Turned over 2,000 lbs of explosives to the 825th Engineer Aviation Battalion, our surplus from the barrier zone.

We have been assigned the job of building the bridge dump based on our estimate of materials submitted previously. We have everything except the steel, which comes by rail.

Personnel — Pfc Baker hospitalized and dropped from rolls.

Marines were reported landing on Iwo Jima, 750 miles from the Japanese mainland.

TUESDAY — 20 FEB 45

Road and bridge maintenance continues. Completed unloading gravel cars; worked 61 civilians today.

Started hauling bridge materials, Esch to Diekirch. Job 20% done; estimate completion 24 Feb 45.

Returned all explosives above basic load to Regimental Supply Officer, the barrier zone being no longer in effect.

Third Army gained up to two and a half miles on its 60-mile front, taking several cities east of the Moselle directly east of Luxembourg City.

WEDNESDAY — 21 FEB 45

Continue bridge and road maintenance. Received no gravel today: worked 63 civilians. Approach road to bridge northeast Ettelbrück being graded; island in this road being filled; land-slide on N7, 1 mi north Colmar, being filled.

Bridge material hauling job 40% complete.

Third Army reached the juncture of the Saar and the Moselle Rivers and has eight miles of Saar River front in the triangle. Organized resistance in this area has ceased.

The failure of mail deliveries is atrocious, and it is interesting to note the Battalion Commander's query to the Adjutant: "Can you even give us a poor excuse that I can pass on to the companies for the failure in mail deliveries?"

THURSDAY — 22 FEB 45

Continue maintenance of bridges and roads. Grading the approach road to the bridge northeast [of] Ettelbrück is continuous. The island in this approach road is 75% filled. The landslide 1 mi north of Colmar on N7 is 50% filled. The 65 civilians working today maintained N7, N8, and N26. The signs on road obstacles for the abandoned barrier zone were removed and picked up 20 Feb 45. Made and reported on reconnaissance of roads and bridges today. Completed hauling all bridge materials from Esch to dump at Diekirch.

Third Army has a substantial bridgehead across the Saar south of Saarburg. The front — 25 miles of the Saar below the Saar-Moselle juncture, five miles from Trier — is held by us. Yesterday 6,000 aircraft hit communications (rail) inside Germany: one of the war's heaviest blows.

Personnel — Pvt Fast is in arrest pending CM for disobedience and disrespect to NCO.[29]

FRIDAY — 23 FEB 45

Continue maintenance of assigned roads and bridges: grading the approach road for bridge northeast [of] Ettelbrück; filled island in that road 90%; completed filling in landslide near Colmar on N7. No gravel arrived today; worked 65 civilians.

Inventoried the bridge materials stockpiled [at] Diekirch. The Company was inspected by the Inspector-General for XII Corps: rating "Unsatisfactory."[30]

Pfc Raymond Fout was not a fast-enough typist to be chosen by S-3 at Regimental Headquarters today, where he went to be interviewed. The Company is not broken hearted over this.

The First and Ninth Armies struck today making numerous crossings of the Roer between Roermund and Duren. Julich is taken. The Third strengthens its Saar bridgehead and is driving across the Saar-Moselle pocket. The Russians have taken Possen.

SATURDAY — 24 FEB 45

Continue maintenance of assigned roads and bridges: grading and graveling approach road to bridge northeast [of] Ettelbrück; filled island in same road 95%; received no gravel; worked 63 civilians.

Reconnoitered new area today, and reported.

Detail removing nails from bridge timber stockpiled at Diekirch.

Hauled 50' D/S Bailey Bridge from Reisdorf to Corps dump, Junglinster. Assigned job of building and draining roads for 4th Armored Division Clearing Station at Metsembourg, north Fischback.

Personnel — Tec 5 Nance, following a long hospitalization in England, rejoined the company today. He says England is a soldier's paradise now, with most of the troops gone.

SUNDAY — 25 FEB 45

Continue maintenance of assigned roads and bridges. Relieved of responsibility for maintaining approach road NE Ettelbrück completely filled the traffic island in the road. No civilians worked today. Received no gravel.

Replacement of Timber Trestle bridge at Christnach under way. The old structure is practically all removed. Rebuilding will start directly the 528th Light Ponton Company completes the Bailey they are putting in there. Materials are loaded and ready to go.

MONDAY — 26 FEB 45

Assigned road and bridge maintenance continues, but with a difference: "Corps still not satisfied with N14, Fels–Diekirch. Though this is in Company 'A' area . . . since 'A' has no civilians, will use some of yours to knock this out to every-one's satisfaction tomorrow. . . . Start work at railroad station 1 mile north Fels grading ruts out of shoulders, backfilling low and soft shoulders with good ballast, and filling holes in road surface. Material for filling can come from railroad at Diekirch or quarry mentioned [in] Message G40. Give this job Hell. Corps lives here!" So — we are doing it. In addition, repairing roads for XII Corps Medical Battalion at Chateau Meysembourg. Worked 63 civilians today, no gravel received.

Christnach bridge — Set abutments and bent: job 40% done. Received following orders: "Have . . . bridge single decked so you can pull Bailey Bridge at 2000 27 Feb 45. Haul Bailey Bridge to Corps dump Junglinster on 28 Feb. Complete Timber Trestle bridge, diagonal second deck, rails etc, 28 Feb."

Reported the stockpile of bridge timbers we found near Chateau Meysembourg: it appears to have been US military dump.

The Army Engineer's office inspected the Company today. Another unsatisfactory. The men are getting sloppy.

New assignment. "Corps reports landslide P987370. Do not know whether mountain has slid on road or road has slid on mountain. If the latter we have shovel loading good material at P916411 which is near Bettendorf on S side river. Fix the slide. Complete tomorrow."

Bittburg is surrounded by Third Army and being shelled.

TUESDAY — 27 *FEB* 45

Continue assigned road and bridge maintenance. Repairing road for XII Corps medics: 60% complete. Working on landslide, bulldozing road clear: 10% done. Worked 61 civilians today. No gravel received.

. . . .

Placed steel, laid first course of deck, will pull Bailey tonight 2000 hours, bridge will be open to traffic in the morning, job 60% done.

Personnel — Pvt Fast was Court-Martialed, found guilty as charged, and restored to duty status from arrest. Tec 5 Willie Barr, injured in Looper's truck accident when we first came to Luxembourg, returned from the hospital and returned to us for Temporary Duty.

WEDNESDAY — 28 *FEB* 45

Continue road and bridge maintenance. The XII Corps Medics' road is 80% complete. The landslide is entirely cleared (mountain off road) and road open to traffic. There is some possibility of a further slide: constant check being maintained. Worked 65 civilians today, received no gravel.

Christnach bridge 100% complete.[31] Traffic was over the bridge 2000 hours last night. A box culvert is being installed adjacent to the bridge, and will be done tomorrow.

Battalion Headquarters notified the Company that the new roads assigned are in Germany, and ours will be the maintenance and repair responsibility of the entire battalion sector.

Personnel — Lt Larew was assigned to the Company today. He and two other new Regimental officers are fresh from the States.

THURSDAY — 1 MAR 45

Completed maintenance work on medics' road for XII Corps at Chateau Meysembourg. On new maintenance assignment for access roads, 101st Evacuation Hospital at Colmar, completed 30% of the work. Worked 94 civilians today on Obersgegen–Wallendorf road. Completed the box culvert next to the Christnach bridge. And after three different sets of orders, hauled the Bailey Bridge from Wallendorf to Junglinster. This is first job in Germany.

The Company is hauling the steel from Valleroy, France, for the seven bridges 1st Battalion has a-building. All of the details seems to be going wrong at once. It is a madhouse, but the steel is getting where it is supposed to be somehow.

Personnel — Bogue, Brehl, Freeman, Gainey, Gamble, Pyke, Rajewski, Rossman, Schara, VanderBroek, and Van Luyn promoted to Pfcs. Pfc Goding made Tec 5. Pvt Fisher went AWOL as of 1800 hrs, failing to return from pass in Lux City.

Lt Larew joined the Company today, and was assigned as platoon commander 1st Platoon.

Third Army is driving for Trier. The First is five miles from Cologne.

FRIDAY — 2 MAR 45

Completed road job for 101st Evacuation Hospital at Colmar. Worked 98 civilians today on N17, N19, and N9. Erected barricade on road over demolished bridge near railroad station Colmar on N26.

Hauling steel from Valleroy continuously. Hauled 963 pieces [of] decking from Esch to dump at Diekirch.

Found one US and four German soldiers dead, unburied, vicinity P889443.

Investigating reported AP [antipersonnel] minefield vicinity Savelborn after checking with 27th AAA Gp OP [Antiaircraft Artillery Group Observation Post].

Personnel — Pvts Kolbusz, Pasquini, and Pisani were hospitalized and dropped from the rolls.

Third Army is clearing Trier. The Russians are reported in Danzig. First Army is about ten miles from Düsseldorf and the Rhine.

SATURDAY — 3 MAR 45

On road maintenance worked 60 civilians on N17 and N19. Hauled Bailey Bridge parts from Mettendorf to Corps forward dump at Messareth railroad station. Unloading bridge timbers and decking at Ettelbrück and hauling to Diekirch: 30% done.

Removed the Anti-Personnel minefield at Savelborn by blowing in place. Field consisted of one row of American M3 charges barely buried and activated with US combination fuze M1 igniters. Trip wires were loose, camouflage of field poor. The field was unmarked. Whoever laid it was sloppy, inexperienced, or both.

"C" Company was ordered today to move into Germany. The Battalion Commander reconnoitered bridge sites over the River Kyll where crossings are being made by assault boat. It is the Third that is crossing the Kyll. Ninth Army has reached the Rhine at Düsseldorf; the First is near Cologne.

SUNDAY — 4 MAR 45

Road maintenance of N14, N17, N19, and Obersgegen–Wallendorf roads continues, even without civilians who neither work nor fight a war on Sunday.

Hauling bridge materials from Ettelbrück to Diekirch 50% complete.

Started removal of Battendorf Bailey Bridge at 1900 hours tonight. Started reconnaissance of mine field which we are going to remove in the entire Battalion area of responsibility. And sent one man to Paris on pass.

Personnel — Tec 5 Oden and Pvt Willis Keller were assigned to and joined the Company today. Pvt Kolbusz returned from hospital.

MONDAY — 5 MAR 45

Competed unloading all bridge timbers on the Ettelbück–Diekirch transfer. Worked 64 civilians today maintaining roads N17 and N19. Removed the Bettendorf Bailey and hauled it to Messerich. Filled the "Y" in the road approaching the Ettelbrück bridge and shaped the road.

Started removal of minefields by clearing eight: 74, 75, 76, 375, 377, 381, 467 (previously removed), 470, and 489.

Personnel — S Sgt Cooke and Sgt Pennington were awarded the Bronze Star Medal today at Regimental Headquarters: SSgt Cooke for his outstanding performance as a platoon sergeant without an officer more often than with an officer and for his excellent work on the Diekirch bridge; Sgt Pennington for his bravery in the mine field on the same bridge. Similar awards were made for Pfcs Dziwenka and Lafferty who were injured in that mine field, and to whom the medals will be forwarded.[32]

Third Army has tactically bridged the Kyll River and is going forward. The Germans report the Ninth Army (US) crossed the Rhine at Düsseldorf. The First is within two miles of Köln.

TUESDAY—6 *MAR* 45

Worked 102 civilians maintaining roads N9 and N14. The additional hauling of decking from Ettelbrick to Diekirch is 65% done. Unloading of gravel at Ettelbrick is 30% done.

Continue clearing mine fields. Lt Larew's platoon is doing this excellent job. Today fields 77, 78, 358, and 518 were cleared. Dead soldiers are being found in these fields in various stages of decomposition: today's yield was 4–6 Americans and 16 [17?—overstrike] Germans vicinity P889459. The entire job is 35% complete.

The Battalion Commander has the Colonel's permission to move Company "B" forward. Billet reconnaissance tomorrow in Germany.

Personnel—Tec 5 Clark was injured clearing the mine fields: his injury is not a serious one.

Fighting is going on in Cologne.

WEDNESDAY—7 *MAR* 45

Worked 104 civilians maintaining N9 and N14.

The assortment of hauling and unloading tasks continues. 80% of the decking going from Ettelbrück to Diekirch is there. Unloaded 8 of 12 gravel cars: 65% done. Unloaded 500 bags of cement at Ettelbrück and hauled it to Diekirch, the Battalion dump across the street from the Battalion Command Post in the Hotel Europa. Started hauling timber for our bridge from Diekirch to Kersch from where it will go forward.

Mine field clearance continues. Cleared today 112, 355, 384, 467, 470, and 494. Another dead German soldier found vicinity P906459.

In Kyllburg Lt Hall reconnoitered the bridge site of our new job where the Bailey is not in yet. The town is not entirely clear of the enemy, but Lt Moseley moved his 3d Platoon there at 1300 today to our first bivouac in Germany.

When Capt Gonzalez was Company "B" Commander, he used to fight the order to name bridges with the very high-sounding declaration: I'll name my bridges when the infantry and the tankers start naming their towns and pillboxes. Now that he is Battalion Commander, he issued the recalcitrant "B" Company an order. "In the past it has been elective with companies whether or not to indicate by a sign by whom the bridge is built. It is no longer elective. You will post each bridge with a legible sign showing the Company and Regiment and giving credit to the Light Equipment Company for participation. It is up to you whether or not you break this down into platoons. Place sign on both ends of bridge. Suggest that the conventional unit designation symbol would make a 'sightly sign.' Post Diekirch, Kopstal, and Christnach before you move."

Third Army spearheads, pushing forward 52 miles from the Kyll bridgeheads in 58 hours, reached the Rhine north of Coblenz. Prisoners of War are being taken in unprecedented numbers.

THURSDAY — 8 MAR 45

Work started today on our first bridge in Germany, at Kyllburg. Abutments and footings for all bents prepared. Job 10% done. Estimate completion for

Lt. Earl E. Hall stands alone on a rock in the river at Kyllburg bridge, preparing to build a crib to support the timber-and-steel bridge under the Bailey span. The Bailey's center support is on rubble, so a crib was built to support the timber trestle. The house to the left of the bridge was quarters for the company. Hall had an office there. Courtesy Earl E. Hall

the 12th. Yesterday, when Lt Moseley wanted to start, he was not allowed to, as the Bailey had not been finished yet.

Road maintenance continues using 60 civilians. Today the civilians worked for us for the last time. 21 of them were given to Company "C" for cleanup work in the billets they left ("C" Company, that is) at the Chateau at Ansemburg.

The assortment of hauling jobs continues apace. Hauling timbers from Diekirch to Kyllburg. Completed unloading the last four cars of gravel. Completed hauling lumber from Ettelbrück to Diekirch.

Cleared minefields 367, 379, 460, 461, 493, 544, and one-half of 488. Found 1 American and about ten German soldiers vicinity P341950.

Third Army has reached the Rhine and the 4th Armored Division corridor is widened to 35 miles. The First [Army] has captured the two-way railroad bridge at Remagen across the Rhine and this almost—miraculously—fortunate bridgehead is being secured. The Nazi sentries who permitted this capture will surely suffer. It is not understandable why this one bridge was not blown, as were all German bridges of any importance.[33]

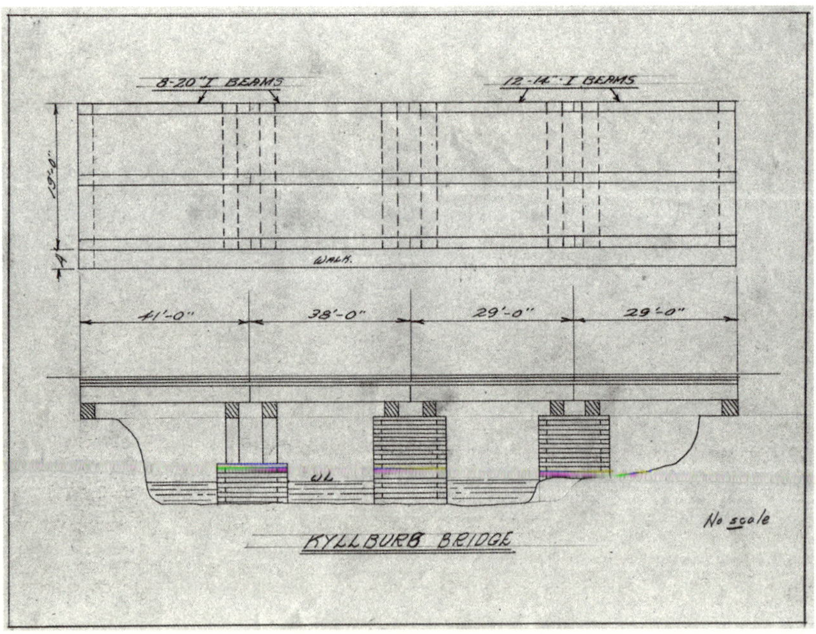

Drawing of the Kyllburg bridge. Courtesy Earl E. Hall

FRIDAY — 9 MAR 45

Completed clearing all assigned mine fields today with the completion of the second half of 488, and 396, 378, [and] 193.

The Kyllburg bridge is 15% completed as building and filling of cribs goes on, and timbers continue to arrive.

The 2d Platoon moved to Kyllburg at 0930, but returned at 1830 unable to find billets in the town. It was expected to move entire Company in the morning. New reconnaissance will be made in the morning to secure living space.

The fleet of Quartermaster trucks attached to us temporarily to haul bridge materials are being relieved of attachment.

First Army's lucky crossing of the Rhine is confirmed and the enemy's opposition is not strong as reinforcements continue to pour over the bridge which is being decked by Engineers to accommodate vehicular traffic.

SATURDAY — 10 MAR 45

Started today to unload the bridge materials being hauled to Kyllburg by Company "A."

The Kyllburg bridge bents 2 and 4 are in place, steel is going into span 4–5. The job is 25% done.

Reconnoitered Kyllburg for billets for Company.

A dead German soldier was found at the mouth of the railroad tunnel where many German civilians were living when the Americans took the town from the Germans. This is across the street from the 3d Platoon living quarters. Besides being a health menace, the cadaver stinks; its removal is being urged.

The Battalion Commander, resorting to his invariable royal third person form of address messaged that "we" understand a pep talk is not really necessary, but "we" nonetheless urge you to hurry up and furnish the bridges that tactically are very necessary.

The news continues good. Three divisions have crossed the Rhine. The Germans are paying through the nose in the pocket formed by Third [and] First Army. The Rhine is being bridged fast: a ponton bridge is going in; ferries are being used to move men and materiel.

SUNDAY — 11 MAR 45

The Kyllburg bridge is 40% complete with all steel set in span 4–5 and the cribbing on bent 3 progressing.

Removed mines from the new site for Supply Point No. 1 at Fliesen.

Having found appropriate billets for the Company, the 2d Platoon moved to Kyllburg today and the Company follows tomorrow. The Battalion Commander is messaging fast and furious these days. "See me or Lt Bubb as soon as possible regarding two more bridges to be built east of Kyllburg. YOU MUST GET YOUR COMPANY MOVED UP THERE WITHOUT DELAY!"

Personnel — Pfc. Baker and Stagner and Pvt Platter were reassigned to and rejoined the Company today.

CHAPTER EIGHT

Germany, Victory, and Frustration

MARCH–JUNE 1945

By the time they entered Germany, the men of Company B were seasoned bridge builders, highway demudders, and scroungers. They had built bridges in and over flooding rivers, under fire, and under heavily traveled temporary bridges. They thought that they knew that the Germans did not fight wars as Americans did. But now they were to see just how differently. Perhaps it is the behavior of a people whose country is invaded, or whose cause is lost in a conventional sense and take up irregular methods. At any rate, the engineers would find the weapons of war stockpiled in places that surprised them, even in an elementary school.

As the war wound down, the men found time for levity. Along the Danube River they discovered an abandoned patrol boat that Major Gonzalez could pilot. The senior officers of the 1303rd enjoyed this toy immensely before reality intruded. Not long after the war in Europe ended, despite the men's hopes and a number of announcements to the contrary, the regiment was ordered to prepare to ship out for the Pacific. Marseilles was to be their port of embarkation directly to the Philippine Islands.

GERMANY

BIVOUAC 22—KYLLBURG, L1861, NORD DU GUERRE ZONE.

ARRIVED 1500 12 MAR 45—DEPARTED 1100 17 MAR 45.

MONDAY—12 MAR 45

The entire Company is present in its first stopping place in Germany, having left the pleasant Chateau of Mersch with its lovely little mistress, Andreé.[1]

Kyllburg bridge span 4–5 completely decked, steel going in to span 1–2, and the job is 55% done.

Oberkail bridge—Cut abutments and removing trees from right-of-way. Job 30% done: estimate completion for 13th.

Road repair—Oberkail–Grosslintgen [Grosslittgen] road, 1 Kilometer southeast Eisenschmitt. Site reconnoitered and work area cleared of mines. The bombed road requires filling of two craters, installation of one culvert and one 45' steel beam bridge. Work 5% done: estimate completion for the 16th Mar.

Personnel—Pfc Gainey was hospitalized, dropped from rolls.

Germany. Courtesy Paul A. Lorah

Eisenschmidt bridge. Courtesy Earl E. Hall

TUESDAY — 13 MAR 45

Kyllburg bridge — Decked span 1–2, set steel span 2–3, expect to complete last (3–4) span tonight and remove Bailey before 0700 hours in the morning. The crews worked all night under Lt Hall when the night before the 1st Platoon failed to do any work and upset the entire schedule. During the night the bridge was completed.

Oberkail bridge — Abutment sills set, approaches cleared, job 50% done but completion date advanced one day. . . .[2]

Road repair — Completed filling the two craters. The pipe for culvert is being assembled: will use two 30" pipes 50' long. . . . The Battalion is preparing to build a 600' bridge at Treis.

WEDNESDAY — 14 MAR 45

Kyllburg bridge — Having complete the bridge during the night, the Bailey was pulled at 1030 hours, some three hours late, and traffic was over the bridge immediately afterwards, with the bridge 90% complete then.

Oberkail bridge — Setting steel: job 75% done. Another day is added to estimation for completion, to the 15th.

Road repair — Installing culvert pipe and hauling fill. Setting steel for the bridge. The job is 45% done on the culvert, 65% done on the bridge. Sketch attached.

BY 14 MARCH the First Army had put five divisions across the Remagen bridgehead up to the Frankfurt autobahn, where hold orders from Eisenhower and Bradley kept them in place. Patton's Third Army attacked in the Eifel region on 3 March (Operation Lumberjack), and in three days the 4th Armored Division had reached the Rhine River. Turning south, Patton pushed his forces between the Rhine to his east and the German First and Seventh Armies to his west, along the Westwall and the Moselle River, threatening to cut off the enemy. On the twenty-second the Third Army crossed the Rhine at two different locations. Several days later the U.S. Seventh Army and French First Army both crossed the Rhine farther south (a follow-on to Operation Undertone). Meanwhile in the north the XXI Army Group launched Operation Plunder, a massive assault to gain a foothold across the Rhine in their sector, supported by the U.S. Ninth Army. Eisenhower intended this effort to be the main line of advance.[3]

THURSDAY — 15 MAR 45

Kyllburg bridge — Completed side rails, cemented base of pier No. 4, and completed the bridge 100%.

Oberkail bridge — Completed setting steel, decking, side rails and end dams; filled and graded the approaches and finished the bridge 100%.

Road repair — Installed culverts and filled over and graded the fill. Completed setting steel, decking, building of side rails and end dams. The job is 100% done.

A cave-in on the side of L78, 1 mile East [of] Kyllburg, was filled 100% today.

Instead of moving forward tomorrow as originally ordered, we will load and haul the Baileys from Kyllburg, Erdorf, and Speicher to Army Engineer Supply Point 18 off N257 south [of] Bittburg.

Third Army attacked across the Moselle. Floating bridges are in, armor is crossing fast. The Battalion has the assignment to put in the first fixed bridge for this attack directly [when] the tactical situation permits at Treis.

FRIDAY — 16 MAR 45

Of the 50-odd truckloads of Bailey which we have to move, about 2–3 have been hauled, clearing Kyllburg and Erdorf. We will clean up Speicher tomorrow.

Grading Corps MSR 257, Daun–Ehlmen–Mayen.

Reconnaissance for the new Command Post is being made with difficulty. If the 10th Infantry vacates Müllenbach, our new Command Post will be there. If not, it will be in the nearby town of Laubach. Both are small villages entirely untouched by the war.

Cleared a mine field of 10 Riegel mines, all booby trapped, by demolition.

BIVOUAC 23 — MÜLLENBACH, L5280, NORD DU GUERRE ZONE.

ARRIVED 1630 17 MAR 45 — DEPARTED 0945 25 MAR 45.

SATURDAY — 17 MAR 45

Moved the entire Company this date, occupying billets just vacated by the 10th Infantry. Trying to keep troops from fraternizing is not going to be easy. And keeping the Germans out of their houses which are earmarked for occupation by us is even more difficult.

Five graders are patrolling the assigned forward MSRs: N257, 259, from Daun, forward.

Personnel — Pvt Hale was injured in a booby-trapped house which he had no business entering. He was hospitalized, dropped from the rolls.

SUNDAY — 18 MAR 45

Graders continue maintenance of MSRs.

Discovered stockpiles of German explosives and ammunition in several locations throughout the village: in the cellar of the school building we are occupying, in a barn, and in a house. Since neither we nor the Security Guard detachment here can handle this contraband, recommended that

Bomb Disposal people pick it along with the cache of rifles turned over to us by the infantry.

Stars and Stripes for the 17th carried a picture of our (the Regiment's) "General Patton Bridge." We were promised enough copies for wholesale distribution, but this promise was not to materialize.

Personnel — Pvt Fast was hospitalized, dropped from rolls.

MONDAY — 19 MAR 45

Switched three graders off the Daun–Uhlmen road and onto L103 and L53 out of Wittlich to Corps left boundary to repair tank damage.

Started hauling bridge timbers and decking from Kyllburg to the Battalion bridge site at Treis: job 20% done.

Started clearing rubble, filling craters, and removing obstructing buildings on the approach road to the Treis bridge. This is a battalion job: "B" Company on the road, "A" and "C" building the bridge, working from both ends towards the middle.

Our planes gave the Germans real hell today: knocked out some 6,000 vehicles trying to flee the Third-Seventh Army trap. The Third's Armor advanced another 15 miles and is now within 15 miles from Mainz am Main.[4]

Personnel — Pfc Brehl placed on Temporary Duty with Headquarters 1st Battalion at the Battalion Commander's request.

TUESDAY — 20 MAR 45

. . . .

Continue work on approach road for Treis bridge: 40% done.

Third Army is moving forward as if on magic boots, catching Germans disorganized and retreating. Mainz is reached, and Worms and Kaiseralautern were taken. The good weather permits plenty of potent air activity in support of the ground forces.

WEDNESDAY — 21 MAR 45

Something new has been added: a program to give every officer and enlisted man in the Regiment a three-day pass.[5] This in the middle of a war seems to indicate the fierce battle expected east of the Rhine is not going to materialize. Anyway, Lt Hall took off with Lt Bubb, the Battalion Adjutant, for three days in Paris, leaving Lt Deveikis acting Company Commander.

Road maintenance with graders discontinued: they cannot improve the roads.

65% of the bridge timbers required at Treis are there. The approach road is 60% done.

Conducting a reconnaissance of all tactical bridging in the Battalion area.

The 11 trucks of decking which Lt Sloat of the 764th, attached to us, is delivering up forward must be intended for a Rhine bridge. He has been gone three days.

The Third drove through Ludwigshaven to the Rhine. Air support continues excellent: communications, airfields, and synthetic oil plants are on the receiving end.

Personnel — Pvt Otterson left the Company for the hospital after being drunk for two days running and showing symptoms of delirium tremens. It is a pity "Pop" had to go this way after wanting so very much to come overseas with us. Too old for induction, he volunteered to fight the war; and when threatened with being kept in the States when we readied for shipment to England, he even falsified his age to come with us.

THURSDAY — 22 *MAR* 45

Completed hauling all bridge timbers required for Treis bridge.

No tactical bridging was found in yesterday's reconnaissance.

One tank is giving a great deal of trouble to Lt Moseley's platoon on the Treis bridge approach road which is 70% done.

Ludwigshaven is completely taken, and there is now only a small gap between the Third and Seventh Armies.

FRIDAY — 23 *MAR* 45

The big news is tremendous. Without any help from SHAEF, Third Army has made the first assault crossing across the Rhine using amphibious tanks and LSIs [LCVPs][6] complete with Navy personnel the crossings were affected last night at 2200 hours against light and moderate resistance. A ponton bridge is in and the bridgehead is being enlarged. XII Corps' Fifth Division did it. Mainz am Main has been cleared.

Continue building the Treis approach road which now is 85% done. Reported the large stores of cloth we found in the Kurhotel und Pension Eifeler Hof, 2d basement, in Kyllburg when higher headquarters requested such reports in order to provide cloth for clothing Germans.

SATURDAY — 24 *MAR* 45

Completed the approach road at Treis bridge in time to open the bridge to traffic late tonight, 24 hours ahead of schedule. This whole job sets some kind of record.

Reconnoitering new Command Post for company.

BIVOUAC 24—GAU BICKELHEIM, WM2037, NORD DU GUERRE ZONE.

ARRIVED 1630 25 MAR 45—DEPARTED 0900 31 MAR 45.

SUNDAY—25 MAR 45

The entire Company moved to the wine-town of Gau Bickelheim, taking Battalion Headquarters along. This is Palm Sunday and all the German people are dressed up to go to church. The weather was fine. We crossed the newly-completed GENERAL MANTON S EDDY BRIDGE at Treis: it looked good. Our Command Post is located about 10 mi East of Bad Kreuznach, about 15 miles from the Rhine. The 3d Platoon was left behind to clean up and move forward tomorrow.

More Paris passes came through: for S Sgt Slade, Sgt Boyes, Tec 4 Russell.

Third Army has taken Darmstadt.

MONDAY—26 MAR 45

Lt Hall returned from Paris, resumed command.

The Battalion Commander left for his leave to London by plane, with the Regimental Executive Officer. The 3d Platoon arrived at 1400 hours.

The roads, as shown by a reconnaissance, are generally in excellent condition, except for several road blocks partially removed to permit one-way traffic. In Mainz am Main the streets are full of rubble making two-way traffic impossible. There are craters and shell holes, too. There was very little traffic.

In the cellar of our billets we found 1400 cases of German canned meat, beef and veal. The stock was reported for medical clearance, but before then the 1st Battalion Companies had their share with which to supplement the operational rations on which we have been subsisting for too long.

Personnel—Pvt Fisher returned to the Company today.

It seems he was not really AWOL, only coming back to us via the Replacement System whence he had been placed by the MPs who picked him up in Lux City on the 1st when he missed the liberty truck. His unsubstantiated story is being checked.

Third Army armored spearheads are in Frankfurt am Main.

TUESDAY—27 MAR 45

Re-decking bridge over tank trap at Sprendlingen, the task assigned, turned out otherwise. Finding the bridge practically gone, the rest of the structure was torn out and the trap filled to make the road passable. Crews will work until job is done.

. . . .

Railroad overpass on N40 leading into Mainz is blown and the dropped steel obstructs traffic. Clearance was started: job 10% done.

Personnel — Sgt Bennett being due for disciplinary action, relieved of duties as Motor Sergeant which were assumed by Sgt Faucett.

Third Army bagged 30,000 prisoners of war today.

WEDNESDAY — 28 *MAR* 45

Completed filling tank trap at Spredlingen 0015 hours this morning.

Continue road maintenance, filling chug holes on L118. Removal of destroyed overpass on N40 south Mainz is 20% complete. Estimated completion of 29th depends upon availability of oxygen as cutting of the steel is necessary before it can be hauled away.

Reconnoitered routes to be cleared in Mainz. Completed road and bridge reconnaissance in the Battalion area.

Third Army has passed Wurtzburg. It is strongly felt that the complete crushing of Germany's tottering lines is only a matter of time.

After a settlement of about 18", topped the fill at Spredlingen.

THURSDAY — 29 *MAR* 45

Personnel — Sgt Bennett reduced to grade of private. Pvt Fisher replaced Pvt Koontz on Temporary Duty at Regimental Headquarters for guard detail. Sgt Gahler placed on Temporary Duty with Regimental Headquarters in connection with the rest center in Luxembourg.

General Patton's armor is again on the rampage, going so far and so fast that contact has been lost.

FRIDAY — 30 *MAR* 45

Road maintenance is urgently needed on L118 between Undenheim and Kongerheim. Bad holes require fill. We cannot do this job not only for a lack of transportation but also because the 1st and 3d Platoons moved to new Command Post vicinity Offenbach. Remainder of the company moves tomorrow. Meanwhile, we are reconnoitering bridging materials for the battalion.

Battalion Headquarters is with Company "A" who are working on the bridge in Frankfurt. Fighting is still going on in the northern part of the city. Frankfurt am Main, a beautiful city, is smashed horribly.

Danzig has fallen. The Third and First Armies have driven beyond Kassel. Armored columns are streaking through central Germany and the Third Reich is tottering.

The Battalion Adjutant left for the hospital today, leaving that Headquarters to the Sergeant Major and the Battalion Construction Foreman, which two staff sergeants are doing excellent work.

BIVOUAC 25—STEINHEIM, M8469, NORD DU GUERRE ZONE.

ARRIVED 1430 31 MAR 45—DEPARTED 0930 8 APR 45.

SATURDAY—31 MAR 45

Moved up the remainder of the Company to Steinheim (near Hanau and Offenbach and Frankfurt) and opened new Command Post 1500 hours.

Hauled last three truckloads of lumber to Company "A" bridge in Frankfurt. Relieved of hauling by new assignment to get ready for occupancy new Third Army Headquarters near Frankfurt.

The Third's armor is again cloaked in security silence, say higher headquarters, but we think it is only because higher headquarters cannot keep contact with the fast moving columns of mighty lightning, and don't know where the tankers are. The First and Ninth threaten to link up in 40 more miles thus isolating the entire Ruhr.

Personnel—Tec 5 Clark was awarded the Purple Heart for wounds received in Luxembourg clearing minefields 6 Mar 45.

IN LATE MARCH the First and Ninth Armies began an effort to encircle the Ruhr region, both commands advancing with great rapidity and heavy losses. German resistance remained stiff despite a lack of fuel. The fighting was localized and remained very intense until late April.[7]

SUNDAY—1 APR 45

On Easter Sunday all the Germans dressed up to go to church and to have a semblance of an Easter Parade in Frankfurt among the crumbling concrete and twisted steel that used to be their beautiful city.

Army Headquarters buildings at Benames found clear of booby traps. We are installing blackout curtains and cleaning up the site. Established maintenance-security guard on floating bridges at Rumpenheim and Muhlheim.

. . . .

Personnel—Pvts Dobrin, Lee, and Murray promoted to Pfc. We found medical supplies and 350 new blankets in the air raid shelter of our Command Post (the school house). We have turned it over to the native Red Cross.

The Third's armor has reappeared fighting in Kassel. The Ruhr is sealed off by First and Ninth Armies: 30,000 PWs. The Third's crossing of the Fulda places us 170 miles from Berlin.

MONDAY—2 APR 45

Completed blacking-out and cleaning, (using Prisoners of War for the latter) Army Headquarters buildings Benames.

. . . .

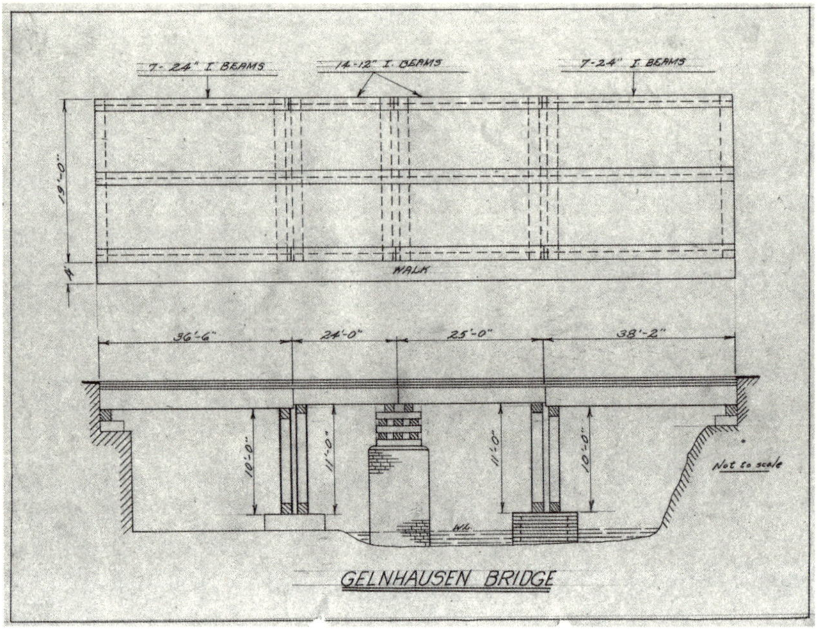

Drawing of the Gelnhausen bridge. Courtesy Earl E. Hall

Yesterday Capt Gonzalez returned by plane from London to resume command of the battalion; Lt Bubb was evacuated to the hospital in France; and Company "C" lost one man, killed by a direct hit of a jet-propelled plane's bomb at their bridge site less than one mile from our Command Post. That plane has been worrying everyone because our Anti-aircraft Artillery can't even touch him.[8]

The war is degenerating. Some 100,000 Germans in the Ruhr pocket are Prisoners of War. Third Army has three armored columns racing towards Berlin: one on the Leipzig road (160 miles from the city); a second east of Fulda 155 miles distant; and a third east of Kassel. Fighting continues in Kassel.

Personnel — Sgt Gahler returned from Temporary Duty at Regimental Headquarters.

TUESDAY — 3 *APR* 45

On Verbal Orders, Army Utilities Officer, constructed two water tank platforms for Army Headquarters at Benames.

On the Rumpenheim and Muhlheim bridges, one machine gun and 6 enlisted men are being established for the night tour of duty, each with a liaison vehicle.

Our old friends, the 217th Anti-Aircraft Artillery Battalion, arrived today and we dug them in about a mile from our area. Their 90 mm guns ought to be able to reach that jet job easily, being able to calibrate the speed of the plane, something the smaller guns cannot do.

Assigned construction of bridge at Gelnhausen.[9]

. . . .

Reconnoitered site, prepared bill of materials, and started hauling timbers. Work 2% done: estimate completion 9 April.

Personnel — Tec 5 Kaczanowski was reassigned to and rejoined the Company. . . .

Regimental Headquarters tells us that after nine months' attachment to the Third Army, the 1303d has been assigned [to the XII Corps], the first such event in history.[10] We have been in the combat zone since 24 Jul without a break.

Kassel has fallen to the Third Army, and they have entered Gotha. One column of armor is within 65 miles of Czechoslovakia. The British have taken Munster.

WEDNESDAY — 4 *APR* 45

Continue maintenance and security guard of Rumpenheim and Muhlheim bridges. The guard, inspected by the Battalion Commander, was jacked up by him quite seriously.

Completed half the road and bridge reconnaissance ordered today.

Gelnhausen bridge is 15% done. One abutment is set, and excavation for the second one is completed. Two of the four bents are being constructed, and steel and timbers are being hauled.

The Russians are five miles from Vienna.

THURSDAY — 5 *APR* 45

Continue guard Rumpenheim and Muhlheim bridges. Warned that these floating bridges are coming out on the 8th, we will guard them even with a rear echelon until relieved by the removing unit: this because on the 8th we move up to vicinity Eisenach, near Bad Salzungen.

Completed road and bridge reconnaissance. . . .

Gelnhausen bridge — . . . Job 25% done.

In its 240 operational days on the Continent, Third Army has captured 240,000 prisoners of war: an average of 1000 each day.

FRIDAY — 6 *APR* 45

S Sgt Sheridan, out reconnoitering roads and bridges, captured two SS Troopers on N276 in a field without a fight.[11]

. . . .

Gelnhausen . . . : job 50% done.

The Esch rest camp closed upon the return of our 1st Platoon from there after a three day period of relaxation there. At the same time Lt Moseley spent three days at Rheims: as his driver he took Pfc McDaniel, and both of them visited near[by] relatives.[12]

SATURDAY — 7 APR 45

. . . .

Gelnhausen bridge — . . . . The bridge will be open to traffic tonight, final completion date 8 April, one day ahead of schedule. Job 90% done.

Reconnaissance performed of new billets and Regimental Headquarters contacted for new assignments at new Regimental Command Post, vicinity Eisenach.

Third Army has captured the German gold supply, worth about 50,000,000 dollars, in a salt mine along with treasures of art. Sgt Faucett was there before the hoard was known to exist, while an Ordnance unit was using the place for a depot.[13]

Gelnhausen timber-trestle bridge underneath a Bailey Bridge, 1945. Courtesy Earl E. Hall

BIVOUAC 26 — FULDA, H3820, NORD DU GUERRE ZONE.

ARRIVED 1430 8 APR 45 — DEPARTED 1000 14 APR 45.

SUNDAY — 8 APR 45

Relieved of bridge guard [at] Rumpenheim and Muhlheim 0800 this morning. Except for the 3d Platoon, left to complete the Gelnhausen bridge, the entire Company moved to Fulda, a former military school beautifully built, apparently for cadets. Warehouses full of rifles, machine guns, and other small arms and hundreds of cavalry sabers and bayonets were found. The souvenir hunters are having a field day.

The new assignments are two bridges: one at Schluchtern, the other at Kerzel. Both replace existing Bailey 80 and 60 feet long, respectively.

The Gelnhausen bridge faulted at bent no. 4 below the water line and settled 18". Final completion will be delayed at least 24 hours, until cause of fault is determined and corrected: estimate new completion date as the 10th.

Personnel — Pfc Rajewski was dropped from rolls, hospital.

THIS BRIDGE AT GELNHAUSEN was a replacement for a steel-girder bridge with two spans, one over water, the other over land. One pier was undamaged and was used in the new structure. When the Bailey Bridge was removed, one bent sank into the stream because the I-beam it rested on was broken. The engineers added rock fill to provide proper support.[14]

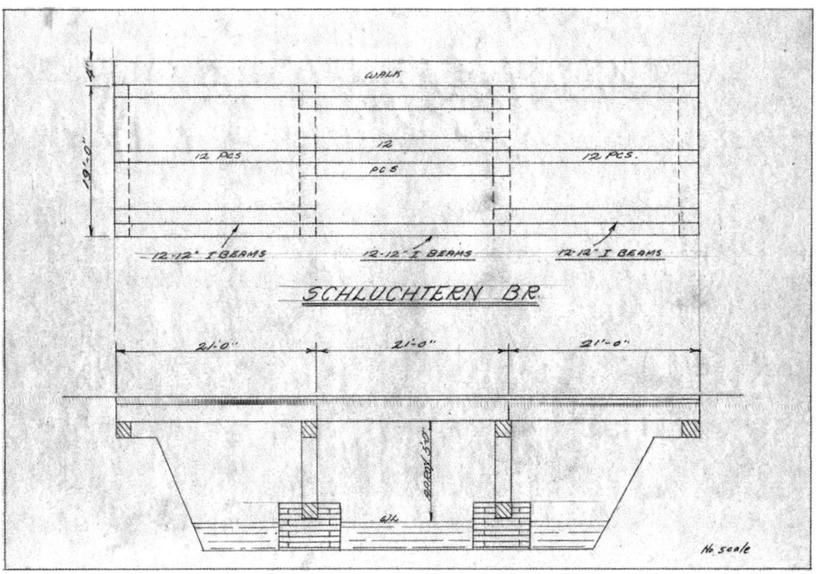

Drawing of the Schluchtern bridge. Courtesy Earl E. Hall

MONDAY — 9 APR 45

Gelnhausen bridge — Removed 60' of Bailey between bents 1–3. Traffic is now traveling on this portion of the new fixed bridge. Bent no. 4 carried the weight of the Bailey for 12 hours without further settlement. The bent has been jacked up to proper elevation and will be allowed to carry the weight of the Bailey and all traffic loads tonight. If it settles no further, the remaining 80' of Bailey will be removed and final work — side rails and curbing — will be completed while traffic is using the bridge.

Schluchtern bridge — . . . . Cleared rubble, cut abutments, hauling decking and timbers. Job 10% done: estimate completion 12 April.[15]

Kerzell bridge — . . . . Old bridge cleared, abutments cut, cribs constructed, cribbing being filled, timbers being hauled. Completed 10% of the job: estimate completion 12 April.[16]

Site No. 679 — Existing bridges will be replaced by one 20-foot span.

TUESDAY — 10 APR 45

Gelnhausen bridge — The questionable foundation was proven safe. The Bailey is entirely removed and the fixed bridge 100% complete. Hauled 60' of the D/S Bailey Bridge to Corps forward dump.

Schluchtern bridge — Set abutments, built and filled cribs, constructed bent no. 3, and hauled all steel. 40% done.

Kerzell bridge — Abutments set, cribs filled, bent constructed, hauling materials. Completion 30%.

No. 679 bridge — Hauling materials: estimate completion 12th. Made road and bridge reconnaissance and reported on it today.

Personnel — Pvt Fisher was hospitalized and dropped from rolls.

The Third Army, still under security blackout, has driven to and is outflanking the city of Erfurt, has captured Coburg, and is less than 50 miles from Czechoslovakia.

WEDNESDAY — 11 APR 45

. . . .

Today we were ordered to haul Bailey from Schluchtern, Kerzell, and Gelnhausen tomorrow, and to be ready to move forward on the 13th or 14th.

THURSDAY — 12 APR 45

. . . .

Company “C” lost their bridge at Hanau when the foundation failed and the bridge collapsed carrying a tank and the retriever. This was not surprising, as the foundation used was the girders of the old structure, partially submerged. The true condition of this insecure footing could not be determined, so no classification was established.

Personnel — Sgt Gahler was transferred in grade to his old outfit the 332d Engineer Special Service Regiment, departed today, and was dropped from rolls. Pfcs James Jones and Gainey and Pvts Beagle and Chalom were reassigned to and rejoined the Company today.

The President of the United States died today.

FRIDAY — 13 APR 45

Trucks hauling Bailey from Schluctern, Kerzell, and Gelnhausen departed 0730 this morning.

The bridge at H105825 100% complete today. Reconnoitering new Command Post vicinity Schleusingen.

Personnel — Pvt Platter was hospitalized and dropped from the rolls today. "A" Co's Lt Everett left for seven days of Information and Education School in Paris.

Nothing could provide a better survey of the amount of entertainment we have had during the war, and the reasons for the sad state of affairs, than the following message from our Battalion Commander written to the special service officer, Lt Nesson: "We appreciate your try and would like to have the show (a USO affair) but under the circumstances at the moment it is impossible. We are absolutely out of transportation. It is 160 miles to Mainz, making it unreasonable to expect troupe to travel that distance tomorrow morning and give two or three performances tomorrow afternoon and evening. 'A' and 'C' companies are still tied up on jobs, making it impractical for more than a handful of their men to attend, and 'B' Company will be on the road that day making it impossible for any of them to attend. Sorry — we just can't handle it." Damnation!

General Eisenhower proclaimed a thirty-day period of mourning for the President. Company "B" in formation observed "Silver Taps"[17] at 2100 hours today for Franklin D Roosevelt.

BIVOUAC 27 — SCHLEUSINGEN, J1316, NORD DU GUERRE ZONE.

ARRIVED 1530 14 APR 45 — DEPARTED 0930 17 APR 45.

SATURDAY — 14 APR 45

Entire company moved to this new Command Post. Ordered to replace 50' Bailey at J220060, near Harras, we reconnoitered the site today.

Mirabile dictu! Twenty-four people of this Regiment are going to the French Riviera for a seven-day leave. Lt Deveikis takes to Battalion Headquarters this evening 1st Sgt Douds, S Sgt Cooke, Sgt Peterson and Pfc Livers and the rest of the 16-man 1st Battalion contingent, and in the morning they will depart for Eisenach where they will emplane for Cannes.

SUNDAY — 15 APR 45

Regimental Headquarters finally ordered "Five minutes of silence will be observed between 2020 and 2025 hours this date to honor of President Roosevelt."

. . . .

MONDAY — 16 APR 45

Harras bridge — Abutments and bent completed, all steel at site, job 65% done, completion date advanced one day, to the 17th.

Personnel — Pvt Koontz's request for transfer to Headquarters Company 4th Tank Destroyer Group was approved, orders issued today, and the soldier departed and was dropped from the rolls.

Harras bridge. Courtesy Earl E. Hall

Third Army has taken Hof, is within seven miles of Czechoslovakia, within 2–1/2 miles of Chemnitz, and less than 40 miles from Dresden. In the Ruhr pocket, the Allies have taken 175,000 Prisoners of War.

BIVOUAC 28 — PRESSIG, J0054, NORD DU GUERRE ZONE.
ARRIVED 1230 18 APR 45 — DEPARTED 1000 20 APR 45.

TUESDAY — 17 APR 45

Harras bridge . . . 100% complete. The Bailey is stockpiled.[18]

The entire Company moved to Pressig today except the 3d Platoon, who will move up tomorrow.

Rothenkirchen bridge — The old structure, reinforced concrete arch bridge, is partially demolished. Repairs will be made with a one-way, one-span steel beam bridge cl 70.

The Regiment was really working these days, witness this message from our Battalion Commander to the Regimental Executive Officer: "Weincek (the Battalion courier) has driven 320 miles today. Please do not send him back unless it is very urgent."

So far 260,000 Prisoners of War have been taken out of the Ruhr pocket. The Ninth Army is in Nürnberg.

WEDNESDAY — 18 APR 45

Hauled the Harras Bailey to Corps dump at Bayreuth.

Rothenkirchen bridge — Abutments prepared and materials located: job 15% done.

Received new assignment: "Open both lanes of Autobahn to two-way traffic with fixed bridge at 077577, Bayreuth Highway A3." Move there directed for tomorrow.

Personnel — Lt Moseley, while taking his 3d Platoon to work, was strafed by a lone enemy plane at Eisfeld about 2100 hours. They were traveling in a four-vehicle convoy behind the light line — consequently using headlights, when the plane came over. Pfc Erickson was wounded slightly in the hand and the attached 676th Air Compressor operator, Tec 5 Cuicci, in the head. Both wounded men were hospitalized, Erickson dropped from the rolls.

Pvt Kreeft relieved of Temporary Duty at Regimental Headquarters and placed on Temporary Duty with the Third European Civil Affairs Regiment, Frankfurt am Main.

Third Army is in Czechoslovakia. The Ninth is in Magdeburg. The First surrounded Leipzig. The Seventh is fighting in Nürnberg.

. . . .

Preparing to use explosives to remove destroyed autobahn bridging, Bayreuth, 1945. Courtesy Dwight Gowdey

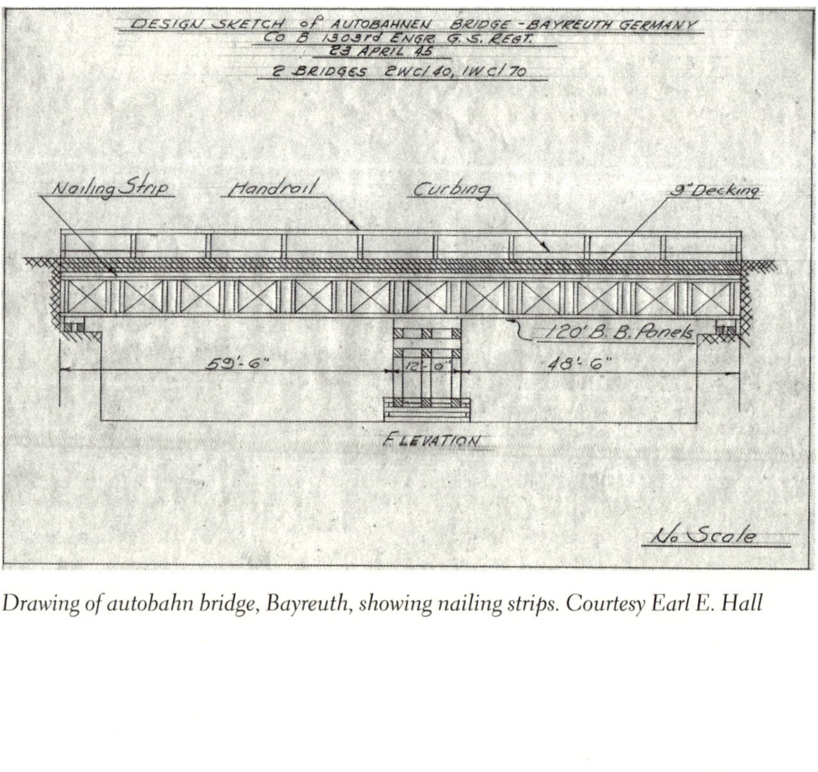

Drawing of autobahn bridge, Bayreuth, showing nailing strips. Courtesy Earl E. Hall

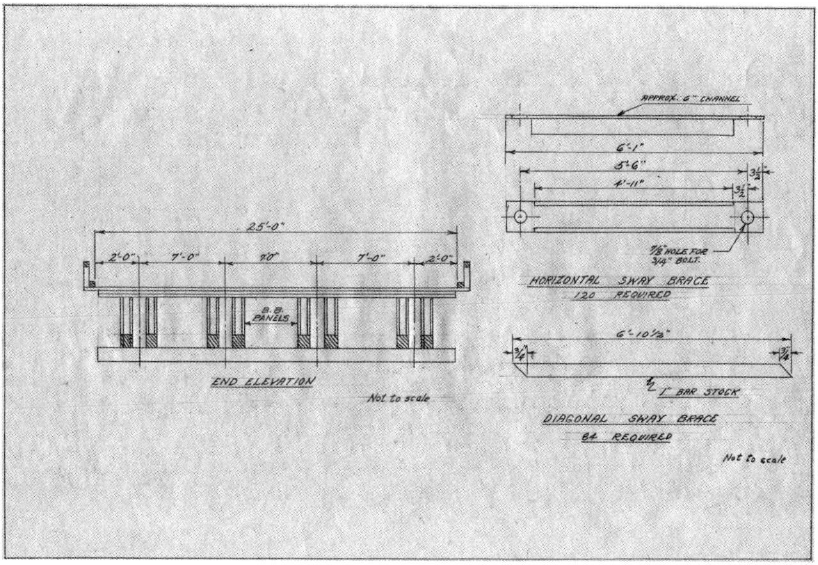

Drawing of autobahn bridge, Bayreuth, end elevation. Courtesy Earl E. Hall

BIVOUAC 29—BAYREUTH, 07658, NORD DU GUERRE ZONE.

ARRIVED 1330 20 APR 45—DEPARTED 0820 1 MAY 45.

FRIDAY—20 *APR* 45

Bayreuth bridge—Blasting of old structure to clear site of rubble progressing. Job 10% done: estimate end 27th. The entire company moved to this location today. We are billeted in the world-famous Wagnerian Opera House[19] which is one of the very few places in town undamaged by bombs and artillery barrages which have ruined virtually everything when this was necessary to convince the reluctant Mayor to surrender.

Nürnberg, the spiritual center of Nazism, is fallen to the heathen Allies.

SATURDAY—21 *APR* 45

Bayreuth bridge—The blasting is being finished. The Bailey panels to be used as built-up stringers and timber are being hauled to site. Completed 15% of the work: advanced final date by one day to the 28th.

Made and reported on reconnaissance of roads and bridges in area.

Personnel—Pfc Erickson was reassigned to and rejoined the Company.

The Russians are reported four miles from Berlin shelling the heart of the German capitol [*sic*]. Less than 50 miles separate the Russians and the

Using Bailey stringers for beams on the autobahn bridge at Bayreuth. Courtesy Dwight Gowdey

Americans. In Czechoslovakia the Third is going strong, and have captured Asch.

SUNDAY — 22 APR 45

Bayreuth bridge — Demolitions completed. Four girders built, timber and decking being hauled to site. Job 20% done.

Again Battalion Headquarters warned us to hurry up, cut the estimated time for completion, and ready two platoons for move on the 27th.

The Russians are in Berlin, and are two miles from its center. The fighting is reported bitter. Our meeting with the Russians is imminent. Seventh Army crossed the Danube yesterday, and is 50 miles from Munich.

MONDAY — 23 APR 45

Bayreuth bridge — Reclassification of the bridge being built. . . .[20] The assignment is really twin bridges one next the other the entire width of the highway. Completion to date 30% with the assembling of two more girders. The cribs are being filled and the bents are under construction. Additional timbers are being furnished from the sawmill we are operating.

Building on the rubble at the autobahn bridge, Bayreuth, with cribbing visible. Courtesy Earl E. Hall

TUESDAY — 24 APR 45

Both abutments are complete the cribs will be filled completely tonight; completion established at 40%.

WEDNESDAY — 25 APR 45

Bayreuth bridge — One pier constructed: job 45% done.

Personnel — Tec 5 Willmeth was reduced to grade of Pvt as part of sentence of Court-martial for drunkenness and insubordination.

Battalion Headquarters and Company "A" moved into Schwandorf, within ten miles of the Czech border. The Russians have completely surrounded Berlin. Third Army is in Regensburg and a column of armor is within 18 miles of the Austrian border.

THURSDAY — 26 APR 45

Bayreuth bridge — One girder is in place. Both bents are completed. Nailing strips are being prepared for the girders and decking is being hauled. The job is 60% done.

Personnel — Pvt Begale placed on Temporary Duty at Regimental Headquarters today.

Third Army crossed the Danube in three places and is closer to Austria. The Seventh is within 40 miles of Munich. The battle of Berlin goes on fiercely.

FRIDAY — 27 *APR* 45

Bayreuth bridge — Placed two more girders for a total of three in place. Continue preparing nailing strips and hauling decking. Job 70% done.

This job is interesting because it is using materials never before used in quite this way: the stringers for the bridge at Bayreuth are Bailey parts. There are four in number, D/S trusses per bridge; a total of eight. The spans are 54'6"–65'6".

Launching is difficult because the girders are limber. They sag and twist to such an extent that they are lying almost flat at the midpoint when still erect at each end. The girders will have to be connected laterally with adequate bracing.

In launching two girders are set up on transoms and using a few rakers and cross braces to accomplish the launching, the girders being launched are separated and pinched sideways into position after they are bearing at three points. This problem was worked out last night finally, but it has delayed completion by at least one day. Completion is moved up two days, to 30th.

Personnel — Battalion Commander wrote to Company Commander: "Reese is being appointed 2d Lt 1400 hours today. Not necessary you be present. Am practically positive he's being assigned to this regiment. If so, he will be further assigned to Company B — in which case he will report to you via Roberson late this afternoon. He is a fine man and I am glad to see this."[21]

Pvt Edmunds was transferred to Headquarters and Service Company and joined them today.

SATURDAY — 28 *APR* 45

Fourth girder in place: job 80% complete.

At a meeting of Company Commanders, the Colonel announced that our work in Europe is just about finished because Third Army's operational mission on the Continent will probably be completed 15 May, the date operations are expected to cease.

Yesterday, Himmler offered to surrender unconditionally to the US and Britain, but not to Russia. His offer was rejected, of course, as the Allied solidarity in [*sic*] unquestionable. In the states there was a celebration which now seems foolish in light of the SHAEF pronouncement that the German offer was not accepted.

SUNDAY — 29 APR 45

Bayreuth bridge — All steel set and all nailing strips are in place. On the east bridge, one layer of decking is on and the second partially complete. This bridge will open to traffic 1000 hours tomorrow. On the second bridge the bracing is being installed, and decking, side rail, and curbing materials are being hauled. Completion to date: 88%.

Personnel — After missing a plane ride from the Riviera — having been outranked on the remaining plane[22] — and a long and tiresome train ride from Cannes to Luxembourg City, and waiting three days for a truck to arrive from the Regiment, Lt Deveikis returned with the five restees from the Riviera today.

The Seventh Army is in Munich. Mussolini has been killed by armed mobsters in attempting to flee with his mistress, and is hanging by his heels in the public square in Milan. To such a pass have the fascistic dictators arrived.

MONDAY — 30 APR 45

Bayreuth bridge — Both bridges 100% complete.

Today we were ordered . . . to move, locating on N20 between Ascha and Parkstettin, there to contact the 101st Evacuation Hospital and start putting in their roads as they are located now in the center of a muddy field; replace with a fixed bridge the Bailey (D/S 50'L) at U608473 and then haul the Bailey to Corps dump at Cham; locate the burned out and shot up US Army

Completed autobahn bridge, Bayreuth. Courtesy Dwight Gowdey

vehicles on the road Parkstettin to Degendorf, get markings, and, if salvageable, evacuate them, if not get them off the road; and place the road along the Danube at U5346 off limits as there are in the river next that spot three barges of some super-duper gas which is toxic even diluted 1:1,000,000 and present in sufficient quantity to be lethal to animal life within a radius of 18 miles on a clear day. Then, there is the usual road and bridge reconnaissance.

Personnel — Pvt Manske was hospitalized and dropped from the rolls.

BIVOUAC 30 — BOGEN, U6346, NORD DU GUERRE ZONE.

ARRIVED 1730 1 MAY 45 — DEPARTED 0920 8 MAY 45.

TUESDAY — 1 MAY 45

Leaving the second platoon to police the bridge site at Bayreuth, the Company moved up to Bogen today, arriving, after a nine hour ride to establish camp in town.

The 101st Evacuation Hospital road will consist of hauling in rock and gravel for laying roads and begins tomorrow.

Ober Altaich bridge — Reconnaissance made. When materials are found, bill of materials and specifications will be submitted.

Road and bridge reconnaissance made and reported today.

WEDNESDAY — 2 MAY 45

Ober Altaich bridge — Two spans will be built using 12" I beams to replace the demolished one-span 20' steel beam bridge. The site is a poor one. Clearing rubble today: job 5% done. Estimate completion 4 May.

Road work for 101st Hospital 10% done.

Schwarzach bridge — Reconnaissance made of site where a fixed steel bridge is to replace the Bailey. Work starts in the morning.[23]

One 45' and the second 50' span will be used. Completion estimated 5 May: work party moved to site.

Because his headquarters could not find billets today and will spend the night on the road, the Battalion Commander spent the night with the Company.

THURSDAY — 3 MAY 45

Ober Altaich bridge — Constructed the crib and filled it. Seated both abutment sills. Constructed and placed the bent. Hauling steel and decking. 50% complete.

Schwarzach bridge — Seated both abutment sills, constructed and partially filled cribs, hauling decking. Design changed: there will be four spans, each 22'6". Work 30% done.

The hospital job is petering out and no work is being done: the field is allowed to dry out. Maybe it is the war, that nobody has yelled? Hitler is reported dead, all of the German armies in northern Italy have surrendered, Hamburg has surrendered to the British. Rundstedt has been captured by the Seventh, and the Third is in Linz, Austria. Prisoners of War by the million are being taken.

FRIDAY — 4 MAY 45

Ober Altaich bridge — Set all steel and laid first course of deck. The Bailey goes out tonight, the second course of decking goes on, and traffic will be over the completed bridge before midnight. At 1800 hours the job stood 85% complete.

Schwarzach bridge — Completed filling the cribs, built and erected two bents, and continued to haul steel and decking. Will set steel on the two spans (end spans) tonight. Job 40% done.

The Battalion is to concentrate in Passau in the next day or two, according to Battalion Headquarters.

Personnel — Pfc Ritchey's application at [*sic*] attend Infantry Officer Candidate School has been approved, and he has been ordered to the school via a Reinforcement [Replacement] Depot prior to jumping into the fire. He departed this date.

At last it is Major Gonzalez. The Colonel came from Regimental Headquarters at Tittling to pin on the golden leaf.

The northern front has entirely collapsed. Everyone is surrendering to the Americans and the English. The two armies in Austria and on the Swiss border have given up to the 6th Army Group. Only the Czechoslovakian border remains active as a front.

SATURDAY — 5 MAY 45

Ober Altaich bridge — The Company Commander forgot to get in to the Command Post in time for Sitrep [Situation Report], so his Lt in writing it followed the plan and reported the bridge only 85% complete when in reality it was 100% complete and traffic was over it before 1800 hours yesterday, meeting the schedule. A change in Regimental Sitrep was requested.

Schwarzach bridge — Existing center pier was cut down two feet to accommodate timber cap. Three of four spans are set and decked. The fourth will be completed and the Bailey removed tonight. Although the bridge will be open to traffic as estimated, completion date is advanced one day to the 6th. Someone caught on! The S-3 wondered, since we had reported no work on the roads for the 101st Evacuation Hospital, were we doing it. Because it would have interfered with the rapid bridging, we abandoned it without fanfare.

The Bailey from Ober Altaich was hauled to Cham today.

The 3d Platoon moved to Passau, at the juncture of the Inn and Danube Rivers to join the rest of the Battalion and to locate new quarters for the Company which is coming up directly, gasoline is available at Supply Point No. 132.

SUNDAY — 6 MAY 45

Schwarzach bridge — All steel set and all spans decked. The 4th Armored Division moved today delaying final completion of the bridge: this will be done tomorrow.

Began road maintenance N85, Schonberg to Passau.

We are immobilized for lack of gasoline which is still unavailable. We will move to Passau when gas is available. The 3d Platoon, meanwhile, continues to hold our billets there.

The Company Commander was promoted today to rank of Captain, along with Lt Bubb, the Battalion Adjutant.[24]

Third Army is pushing in Czechoslovakia and has taken Pilsen and is heading for Prague. A restraining line, however, is holding them back: they cannot go beyond it because war's end is expected any time now.

MONDAY — 7 MAY 45

Schwarzach bridge is 100% complete today.

Reconnoitered bridge site Neufelden [actually Rottenegg], Austria, prior to replacing the 120' Bailey.

Personnel — Sgt Cunningham ["dropped from rolls to hospital" on 19 April 45], Tec 5 Kawula ["hospitalized and dropped from the rolls" on 3 April] and Pvt Hale [injured in a booby-trapped house on 17 March 1945 and hospitalized] were reassigned to and rejoined the Company today. New men assigned to and joined were, Pvts Adams, Boardman, Boyd, Brigman, and Harlan Bullock.

Gas was finally obtained today: tomorrow we move.

BIVOUAC 31 — PASSAU, XQ2113, NORD DU GUERRE ZONE.

ARRIVED 1300 8 MAY 45 — DEPARTED 0615 1 JUN 45.

TUESDAY — 8 MAY 45

V-E Day! The European war is over officially. At 0241 yesterday, 7 May, the German High Command surrendered unconditionally to the Allies at Rheims. Hostilities cease tomorrow, 9 May 45.

Three hours after leaving Bogen, we opened the new Command Post in Passau, just as Lt Moseley's platoon left for Rottenegg, Austria, erroneously called Neufelden yesterday, to build the last bridge.

Some one forgot to tell the Germans the war is over. The bastards strafed us in town with three passes down the main street. It was a lone plane: soldiers and civilians alike were frightened.

Personnel — Pvt Shourd was dropped from rolls to hospital. That old AWOL status of Pvt Fisher's was finally cleared up today: he was not AWOL. The M/Rs [morning reports] have been corrected accordingly.

If we remain on the Continent for any length of time, we will be XII Corps troops.

WEDNESDAY — 9 MAY 45

Rottenegg bridge — . . . .[25]

The stringers will rest on double bents in the center of the old bridge and on standard abutments at either end. Both sills seated, footings for bents prepared, and timber being hauled in. Estimate completion for the 12th: complete to date 20%.

Hauling Bailey (D/S 100'L) from Schwarzach to Cham.

Personnel. — Pfc Kruk was reduced to Pvt by court-martial and then dropped from rolls following hospitalization.

All fighting has ceased except for small sections near Prague.

Last bridge the company built — a postcard view of the original bridge. Courtesy Kathy R. Hart

THURSDAY — 10 MAY 45

. . . .

Personnel — Pvt Kemper was transferred in from Headquarters and Service Company, and Pvt Chalom was dropped from rolls after hospitalization.

FRIDAY — 11 MAY 45

Rottenegg bridge — No change in completion as 3d Platoon continues to wait on steel.

The Military Government Detachment at Passau cannot handle the security guard (which will be taken over by the 5th Division when they get here). The Displaced Persons [known as DPs] center in town — rather, the Russians in that center — are looting Passau, smashing windows and getting drunk on stolen liquor and wine. The Battalion was requested to furnish guard. One platoon from each of the three companies will be this guard. A roving patrol with machine guns mounted on trucks will cover the town, and a fixed machine gun guard at the center will keep the DPs in there. Orders are to shoot. After an hour's visit to the Center, it is understandable. Yet it does seem strange, to have to have more trouble with the nationals of one's Allies than with the vanquished foe.[26]

Personnel — Pvts Glenn and Johnson dropped from rolls, hospitalized.

At an officers' dinner at Regimental Headquarters (Company "D" Command Post) in Degendorf the 1303d entertained some visitors from the Engineer sections of XII Corps and Third Army. They in turn told us what a good bunch of boys we were. There was much mutual admiration and slapping of the back, a good steak dinner, and wine, beer and even Scotch to help the speakers along.

SATURDAY — 12 MAY 45

Rottenegg bridge — No steel, no work. Completion date advanced two days, to the 14th.

Passau Security Guard continues unchanged, except that the Russians are great respecters of machine guns. There was no trouble all night.

Personnel — Pvt Kruk was reassigned to and rejoined the Company. S Sgt Dunbar and Tec 5 Maas departed for seven day furloughs on the Riviera.

The Germans are fleeing their Russian captors, and even whole Panzer divisions are trying to surrender to the Americans instead of the Russians. Vehicles by the thousands and Prisoners of War by the tens of thousands are pouring towards our lines.

The regiment's V-E Day dinner, Major McEnary (3rd Armored Division or XII Corps) speaking. To McEnary's right, Colonel Pruetl and Colonel Vandenberg; to his left, Major Gonzalez, Major Gradian, and Captain Denman. Courtesy Dwight Gowdey

SUNDAY — 13 MAY 45

Rottenegg bridge — Steel began arriving late this afternoon, and the setting of steel began then, too. Job 65% done. Security guard, Passau, continues.

Personnel — Pfc Williams was reduced to grade of Private.

MONDAY — 14 MAY 45

Rottenegg bridge — Steel all set, nailing strips on, and decking begun. Job 75% complete. Completion date advanced another day.

Security guard continues uneventful.

Personnel — Pvt Chalom was reassigned to and rejoined the Company. In order to control the surrendering Wehrmacht, barriers have been established beyond which they cannot flee in their attempt to evade the Russians who are out for German blood, and revenge. Should any Prisoners of War pass these lines, the Americans will return them to Russian controlled areas.

. . . .

WEDNESDAY — 16 MAY 45

The 3d Platoon returned to Command Post from Austria where they completed the last of our European bridges and the first one Capt Hall named — "Ivan — East End of the Line."

Security guard, Passau, continues.

Battalion Headquarters is all adither about a "Sissy Britches III," a ship which has been confiscated and which the Colonel is giving to Colonel Dougherty, XII Corps Engineer.

THURSDAY — 17 MAY 45

Continue Passau Security Guard.

There was a Red Cross Doughnutmobile here today.

The Battalion was talked to by the Regimental Commander in the Niebelungen Hall. He told us that he had seen some troop lists which did not contain our name for redeployment direct to the Pacific; that we had been assigned to XII Corps and would wear their shoulder patch as well as probably work for them for some time; and that even if we did go to the Pacific theater our chances of going there through the States were excellent. Everyone felt very well after these announcements.

After the meeting the three Company Commanders accompanied the Battalion Commander who accompanied the Regimental Commander to Regimental Headquarters via the Danube on "that" boat.[27]

FRIDAY — 18 MAY 45

Continue maintaining the Security Guard in Passau. Lt Larew, Commander of that guard and a vital implement of the enforcement department of Military Government, is having a high time arresting offending Germans for violating curfew regulations and is also finding some American soldiers in bed with people of "verboten" categories even if they do say they are refugees and simply displaced persons. The routine for this is for a soldier to say to a husky wench, "You're Czechoslovakian, aren't you?" and the inevitable answer is "Jawohl."

Large numbers of German Prisoners of War who came to the American lines from Russian-occupied territories have been turned back to the Russians: 35,000 alone from the 11th Armored Division.

Show-down inspections are in the offing for individual and organizational equipment. And six men from the Regiment, all old timers and high point men, are scheduled for the first contingent of troops to be rushed back to the States now that the war's over. Included in this group are "A" Company's 1st Sgt Butts and our own Sgt Lamb.

SATURDAY — 19 MAY 45

Security guard continues.

Sgt Lamb departed for Regimental Headquarters and the States today.

The Information and Education program is going ahead full blast, with a Regimental School established on paper, schedules published and instructors selected.

SUNDAY — 20 MAY 45

It being Sunday, a day of rest, we got a job: filled the drainage ditch for road to lumber camp of the 1137th Engineer Combat Group.

The security guard alone continues.

MONDAY — 21 MAY 45

Passau guard continues.

Censorship has been discontinued, to be restored only when a unit is alerted for shipment. Cheers!

Personnel — Pfc Cederberg, Pvts Fisher and Fast were reassigned to and rejoined the Company today.

TUESDAY — 22 MAY 45

Passau security guard taken over by Lt Reese since Lt Larew is going on leave to Paris with Lt Stoudt of Company "C" and Lt Laue of Company "A." Lt Kessler of "A" goes to London. And to Paris goes S Sgt Holm, Battalion Sergeant Major. A Regimental History is up.

Personnel — Sgt Lamb departed today, transferred to Ft Sheridan, Illinois, USA.

WEDNESDAY — 23 MAY 45

Passau security guard continues.

Personnel — Pfcs Gryder and Wright hospitalized, dropped.

Talk is heard of concentrating the Regiment at Regensburg, then at Degendorf.

THURSDAY — 24 MAY 45

Last night trucks hauled DPs [Displaced Persons] for Military Government.

A last big bridge job has reared its ugly and unpeaceful head. Reconnaissance was made today for billets vicinity Regensburg where not much is available because it is Corps Hq. We plan to move tomorrow.

The 5th Division MPs relieved us of Passau Security Guard at 1900 hours.

Personnel — Pvt Kreeft, on Temporary Duty with Military Government, relieved of assignment to the Regiment and transferred to 12th Army Group today.

FRIDAY — 25 MAY 45

Two reconnaissance parties continue hunting billets in the vicinity of Regensburg. Because Corps Headquarters is there, even if space were available we could not move in. In addition to a happy home, these reconnaissance parties are trying to find materials with which to build the bridge at Regensburg — a major task. The plan now is to move the Company at 0830 tomorrow morning.

Personnel — Pvt Hale was hospitalized, dropped from the rolls today. And Pfc Kelly reported to Regimental Headquarters for Temporary Duty as one of the members of an enlisted committee recommended by the companies, to write the Regimental History.

SATURDAY — 26 MAY 45

Continue reconnaissance of bridging materials.

Move one platoon to vicinity Regensburg as a billeting party. Tomorrow the balance of the Company will leave Passau. This was the plan in the morning.

In the afternoon censorship was clamped on again, and we were alerted for shipment direct to the Pacific Theater of Operations.[28] All equipment is to accompany us and we are to be ready to depart for the staging area by the 15th of June. The probable departure date from the Continent is June 25th.

The reaction of the men was not good, as they had all got the idea from the Colonel's talk same days ago that we would at least ship through the States in going to the second theater of war for duty. Although they should not have got this idea, as it was not promulgated, there's no explaining their own mistake to them. The platoon in vicinity Regensburg will move back in the morning.

SUNDAY — 27 MAY 45

Preparations are being made for showdowns and inspections. The 3d Platoon returned from Regensburg. Lt Reese left for a three-day Paris leave. The 764th Engineer Dump Truck Company and the 676 Engineer Light Equipment Company were relieved from attachment to the Regiment.

Personnel — S Sgt Dunbar and Tec 5 Maas returned from furlough on the

Riviera. Pfc Kelly returned from Temporary Duty at Regimental Headquarters, where the Regimental History was abandoned even though no work had been done on it.

. . . .

TUESDAY — 29 MAY 45

Began building crates. The 32d Ordnance inspection team checked all T/E [Table of Equipment] weapons.

Personnel — Pfc Dziwenka was reassigned to and rejoined the Company. His Bronze Star Medal for the wounds of Diekirch was finally personally presented. Also reassigned and rejoined were Pfc Rajewski and Pvts Glenn, Platter.

WEDNESDAY — 30 MAY 45

Continue construction of crates. Our last Red Cross Doughnutmobile visited us today. After a really hectic rat race and half a dozen meetings, the plans were changed. Now we depart Passau for the marshaling area on June 1st.

Personnel — Pfc Wright rejoined the Company from hospital.

THURSDAY — 31 MAY 45

Packed the Company for shipment via rail to the staging and marshaling area at Marseilles, France. The trucks, carrying all T/E [Table of Equipment] allowances except mess equipment, go separately in a Regimental Convoy under Lt Moseley. The mess truck will deliver equipment and personnel to Regensburg siding to load and be ready for us tomorrow morning. We will move early in 764th trucks to Regensburg, there entrain and take off for Marseilles.

The Regimental Headquarters guard, long on Temporary Duty, returned to duty: Pvts Begale, Gavinsky, and Puccio.

ENROUTE TO CALAS STAGING AREA — BY RAILROAD

ENTRAINED 1130 1 JUN 45 — DETRAINED 2215 6 JUN 45.

FRIDAY — 1 JUN 45 TO WEDNESDAY — 6 JUN 45

At 0615 hours we departed Passau, Germany, by motor convoy enroute to Regensburg there to entrain. The hauling was done by the 764th Dump Truck Company, this being Lt Shaffer's last assignment from the Regiment. We entrained at 1130 hours but the train did not leave Regensburg until 1600 hours.

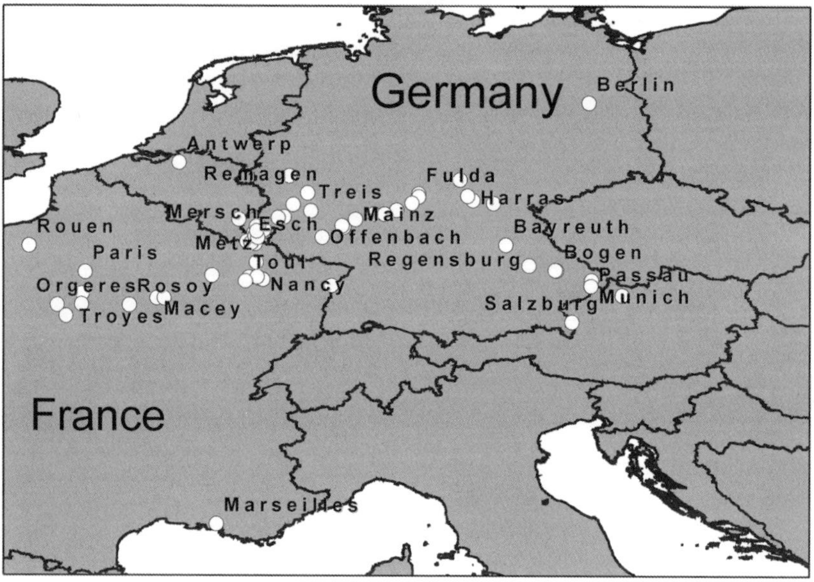

The End of European Service. Courtesy Paul A. Lorah

Personnel—Cpl. Hosking was transferred at the train and departed for the 1135th Engineer Combat Group from where he will be shipped to the States, being overage. 29 enlisted men were promoted to Pfc: Blagun, Chalom, Dzwerzynski, Feuer, Gavinsky, Golden, Gossett, Gruppi, Hale, Hughee, Kanachey, Keller, Kepka, Kolbusz, Kruk, Kustera, Levey, McDaniel, Michek, Niczay, Pasquini, Pisani, Rosinski, Sharpe, Strobel, Williams, Willmeth, Woodward, and Zins.

After the first, the trip settled down to a long, tiresome, uncomfortable, and protracted affair. 24 men were living in a car with only straw to sleep on; besides, the chevaux had been there. We ate well, though, out of our own kitchen cars, three meals a day. Company "E" suffered the only casualties when their kitchen caught fire. One man was killed jumping out of the inferno, several others were hurt, and all their equipment was destroyed. The boon of the trip was the good weather.

Arriving at Marseilles, France, at 2200 hours the night of the 6th, we entrucked and were taken out to our bivouac in the Calas Staging Area which we found only with a great deal of difficulty as the advance party did not expect us until the morning and the static staff did not seem too certain of the location of the area which they had assigned us. It wasn't until after midnight that we found our tent city and bedded down for the night.[29]

MARSEILLES, FRANCE

CALAS STAGING AREA

ARRIVED 0100 7 JUN 45—DEPARTED 0700 23 JUN 45.

THURSDAY—7 JUN 45

The beginning of our POM [Preparation for Overseas Movement] this time was just as inauspicious as the others had been. We were apparently not expected so it took time to find our way out to the area. Once there and finding our tents up and the motor vehicle convoy asleep, we slept.

The morning saw the beginning of the inevitable POM rat race. At 1430 hours the Company in formation heard the 28th Article of War pertaining to desertion and shirking hazardous duty. The staging area is not completed yet, newspaper accounts in the States notwithstanding. Mess buildings are in process of construction; sewage disposal facilities are only now being installed; many of the services so ambitiously planned and stridently reported in the press are still causing things such as three-and-four-hour lines for beer at the "GI Beer Garden." The way to beat this is to sweat out one's tickets one day and get the beer the next. There is a line for everything.

The static staff is augmented by German Prisoners of War who, among other things, furnish the KP ["kitchen police"] details which are markedly efficient. Whatever else may be said of them, the Germans know how to work.

Boxes are being built, equipment is being processed—either replaced totally, as was the kitchen equipment, or salvaged. Summer clothing is being issued with difficulty due to lack of sizes and that peculiar inability of the Supply Sergeant, Jenkins, to get the right quantities of the correct items in proper sizes. Ordnance and Engineer services are processing their own issue equipment.

SATURDAY—9 JUN 45

Tec 5 Howery replaced Hosking as a Cpl and Pfc Vamer was promoted in the former's place.

SUNDAY—10 JUN 45

Cpl Groom was hospitalized today and Pfc Mannion replaced him on Temporary Duty at Regimental Headquarters as Company Clerk.

WEDNESDAY—13 JUN 45

That fall Lt Larew had on the Bayreuth bridge continued to bother his back so he was hospitalized today for observation and treatment.

THURSDAY — 14 JUN 45

Cpls Howery and Jenkins, Tecs 5 Kawula and Ray, and Pfc Dziwenka were all transferred to CAS Detachment of the Staging Area for shipment to the States on points. They will not come with us overseas.

FRIDAY — 15 JUN 45

Replacement promotions were made today. Sgt Bennett is a Sergeant again, Cooper was appointed Cpl, and Mosier and Rajewski made Tec 5.

The chief recreation when POM work is suspended for the night is passes to Marseilles and Aix. The famous port city is about 18 miles away and most of it is "Off Limits" to American and British troops. When the transportation we had was taken away for processing, the difficulty of getting into town was almost — but not quite — insurmountable. The day we arrived here the much touted riot between the Senegalese and our own Negro troops took place, and armed guards are still posted all over town. The restrictions placed on the bawdy houses stem from the terrific VD [venereal disease] rate and have driven the prostitutes into the streets where they solicit. Before the native quarter was placed out of bounds some American soldiers were being killed there nightly either for their money or for hatred. But soldiers have an aversion for observing signs. Even First Sergeants sometimes have flesh weaker than spirit and end up by spending the night in the local jail. The flesh pots are definitely here and people are reveling in them.

SATURDAY — 16 JUN 45

Pfc Kanachey was hospitalized on the 13th and dropped from the rolls today.

MONDAY — 18 JUN 45

Pfc Kerkin got back his Cpl's stripes today after many months of good service following a reduction based on no T/O [Table of Organization] position vacancy.

THURSDAY — 19 JUN 45

S Sgt Slade, Sgt Peterson, and Cpl DeMattei, our three "old men," were transferred to the 5th Labor Supervision Battalion to sweat out their remaining days in the European Theater of Operations before being shipped home. Pfc Brehl returned from Temporary Duty with Battalion Headquarters.

There were promotions all the way up the line. Boling got his Staff Sergeancy and replaced Slade as Lt Moseley's sergeant. Chavez is a sergeant as is Good; Anselmo, Hetler, and Rajewski are Corporals; and Lange and Jessie Bullock Tecs 5.

WEDNESDAY — 20 JUN 45

We got replacements today: Sgt Hasenkamp, Cpl Chrisotpher, Tecs 5 Foglio, Francis, Giambrone, and Hart, Pfcs Ormsby, Paffie, Perrine, Charles Smith, Chester White, Chase, and Diez, and Pvts See, Messner, Ripley, Franklin Smith, Howard Smith, Soares, Westbrook, and Wojewodski.

Pfc Garcia was hospitalized, and Lt Larew returned from the hospital for duty.

THRUSDAY — 21 JUN 45

Cpl Groom returned from the hospital for duty.

FRIDAY — 22 JUN 45

The final European Theater of Operations Morning Report today dropped Pvt Fast and picked up Garcia.

CHAPTER NINE

On to Japan

JUNE–NOVEMBER 1945

The men of the 1303rd were not appreciative of the honor implicit in their being chosen for high-speed transfer to the Pacific theater. Nonetheless, behind this movement was the recognition that their unit was a good veteran regiment. Only three general service regiments served in the combat zone in Europe, all in Third Army: the 1301st, 1303rd, and 1306th EGSRs, sister regiments that had trained together at Camp Ellis. Poor as they thought their instruction at Ellis was, it obviously was good enough.

The 1303rd was to land on Kyushu (in Operation Olympic, the first phase of the planned invasion of the Japanese home islands, Operation Downfall) on D+1 to build airfields. Yet Earl Hall is firmly convinced that they were to go ashore to clear the beaches in support of the 1st Marine Division's landing. Regardless, the engineers would learn during their Pacific service that fresh pineapple was really good and Filipino "rot gut" was very dangerous stuff. They also very much appreciated the cumulative effect of the use of nuclear weapons at Hiroshima and Nagasaki, completely convinced that these strikes alone brought about the Japanese surrender and saved many of their own lives. In the more rarefied reaches of the Pacific commands and the Pentagon, a debate was just starting on the wisdom of invading Japan, especially with the updated data on defenses provided by MAGIC (from intercepted Japanese radio communications). In the postwar years this debate — intertwined with arguments over the use of the atomic bombs — has grown. Company B's diary reflects the opinions of the men and of most veterans who had expected to invade Japan.[1]

While many veterans of the European theater, including Generals Eisenhower and Bradley, dismissed the ground combat in the Pacific as "bush league stuff," it was, if anything, more vicious than the fighting in Europe.[2] Many of the men of the Thirteen Third shared in this assessment and were fortunate never to learn the truth the hard way. The reason they and other men were being prepared to transfer from Europe, many by way of in-transit leaves in the United States, was to support the invasion of Japan. Based on experience, U.S. planners as well as Pres. Harry Truman, expected to need about 600,000 men for the first stage of the invasion and to suffer a million or more American casualties (killed and wounded) in the entire operation.

AT SEA

ABOARD USS GENERAL JOHN POPE

EMBARKED 1000 23 JUN 45—DEBARKED 1000 27 JUL 45.

SATURDAY—23 JUN 45 TO FRIDAY—27 JUL 45

Very early this morning we began preparations for departure from the European Continent. Breakfast was at 0400. By seven o'clock the tents were empty, the area policed, and we loaded on trucks where we sat for perhaps an hour before leaving the Calais Staging Area for the Marseilles Port of Embarkation.

Embarkation was the usual long-drawn out process beginning, for us, about 0930 hours and ending at 1000 hours. In the process the field desk which was being carried as baggage became lost and Tec 5 Reimer and Pvt Willmeth had a merry chase of several hours duration before they recovered it. There were doughnuts and lemonade furnished by the Red Cross, and even a Navy band that played occasionally.

Our vessel is a navy transport, USS General John Pope. It's carrying besides our regiment the 1306th Engineer General Service Regiment, a combat battalion of engineers, two MP battalions, an Engineer shop company and truck company, and some quartermaster troops. We are the second contingent of troops to be redeployed direct from Europe to the Pacific Theater.

The Regiment is working on the ship. This is good as it breaks the monotony of a long journey and gives the working men a chance to buy cokes and ice cream which the rest of us do not get. The food is Navy food and is better than anything we have eaten aboard a ship yet. The officers and men who are working eat three meals a day the non-working passengers eat twice.

We sailed from Marseilles, France, at 1630 Sat 23 Jun. Present were 5 officers, 172 enlisted men: T/O plus authorized 2% overstrength.

SUNDAY—24 JUN 45

Pfc Chalom hospitalized in ship's sick bay.

MONDAY—25 JUN 45

We passed the Rock of Gibraltar about 0100 hours this morning. It could be seen dimly through the haze.

WEDNESDAY — 27 JUN 45

Pfc Chalom returned from sick bay to duty.

SUNDAY — 1 JUL 45

In the Caribbean Sea.

MONDAY — 2 JUL 45

Tec 5 Maas went into sick bay today.

BALBOA, PANAMA CANAL ZONE

DEBARKED 2200 3 JUL 45—EMBARKED 0100 4 JUL 45.

TUESDAY — 3 JUL 45

This morning we entered the Panama Canal Zone. The Canal itself was negotiated during the afternoon[3] and at 2000 hours we tied up at Balboa for the night. The ship's crew and passengers were permitted ashore between the hours 2200 to 0100, in two shifts, to see a USO show, get some doughnuts and punch, buy candy and tobacco and Panamanian souvenirs (those that could afford the latter) and magazines and ice cold cokes in unlimited quantities.

Of all the sights, the most notable was the MP shine. They were positively beautiful. Schooled in the peacetime garrisons of the Regular Army, their brass glittered, their leather gleamed, their tailored uniforms fitted "with ne'er a wrinkle," not unlike Noyes' Highwayman's. Though we got off only on the dock, it was good to feel solid earth again. That the thousands of men who throughout the European war had lived for the most part like pigs should have comments on the beautiful Panama MPs is natural. It was good to see soldiers looking like soldiers.

Someone bought a *Saturday Evening Post,* issue of 23 Jun. Lo and behold! we were practically famous. In the leading article — the "Patton Legend" — was a picture of the General standing before the "General Patton Bridge" built by the Mighty Midgets (Co "F") of the 1303d Engineers. The remainder of the trip provided an opportunity for us to poke this article at the noses of everyone else, especially the less fortunate 1306th.

PACIFIC OCEAN

WEDNESDAY — 4 JUL 45

We sailed again from Panama late in the morning and proceeded to our next stop, several days hence, in the Marshal Islands.

THURSDAY — 5 JUL 45

Tec 5 Maas returned from sick bay for duty, and Pfc James Wilson went in.

WEDNESDAY — 11 JUL 45

Pfc James Wilson returned to duty status from hospital.

The comfort one would like on a ship in tropical waters is not here on this overcrowded transport. That, of course, is in the nature of things. The complement being all male, nearly everyone is running around in shorts getting tanned. Everywhere at least two soldiers congregate, there is bitching. Some of it is not without some justification. The movies, for instance. Newspapers say transports show movies daily. That is true. But accommodations restrict audiences. Company "B" saw two on the entire trip. Everything else is about the same way, except Orientation which comes every day. The showers — salt water, to be sure — were a boon, as was a good ship's store, this time a regular Army PX [post exchange]: everything was sold equitably without the usual discrimination in favor of the Navy. The men all enjoyed the fresh fruit which provided a bite between meals. The trip continues apace, its monotony broken only by an incredible amount of gambling.[4]

WEDNESDAY — 18 JUL 45

Crossed the International Date Line losing a day.

FRIDAY — 20 JUL 45

We paused at Eniwetok in the Marshal Islands for fuel. Not since the Normandy beaches have we seen such a concentration of Allied Shipping and Naval might as there was here. That is a most stirring sight.

THURSDAY — 26 JUL 45

In Philippine waters today, everyone who had rumored our arrival at such places as Okinawa is slightly miffed. Too bad.

FRIDAY — 27 JUL 45

Today we steamed into Manila Bay where we are spending the afternoon and night amid the many, many other ships there.

PHILIPPINE ISLANDS

CALUBANG, LUZON

ARRIVED 1600 28 JUL 45 — DEPARTED 1100 1 AUG 45.

SATURDAY — 28 JUL 45

Debarkation started this morning at 0800. The Company's first LCM [Landing Craft, Medium] landed on Manila's battered shores about 0900 hours. By 1000 hours the entire Company was ashore. We entrained at the

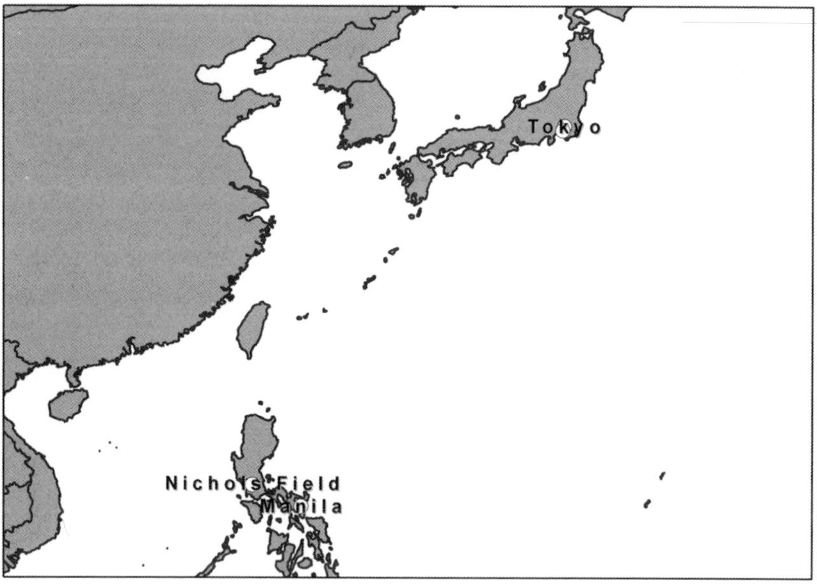

The Philippines and Japan. Courtesy Paul A. Lorah

port and after a three hour ride — box cars again, or, rather, canvas-topped gondola cars — we arrived at an established transient camp where we became attached to 121st Chemical Processing Company for rations and quarters for our stay here. The first day in the Islands started with purchases of pineapples, coconuts, and bananas, and the native rot gut against which the entire Company was warned this afternoon.

SUNDAY — 29 JUL 45

There is a Regimental Mess staffed by the companies each day. The food is not bad. There is no work while our area is being selected on Nichols Field.

MONDAY — 30 JUL 45

Three men were hospitalized today, Woodward, Lee, and Fisher. The latter is a poison liquor case, and is reported blinded.

TUESDAY — 31 JUL 45

Tecs 5 Nance and Whitaker were hospitalized today. Yesterday's report on Fisher is confirmed: he is totally blind and will be returned home.

NICHOLS FIELD (MANILA, LUZON)

ARRIVED 1700 1 AUG 45—DEPARTED 1400 27 AUG 45.

WEDNESDAY—1 AUG 45

The transient camp at the small village of Calubang is being deactivated and so today we were kicked out by the outfit running it. An advance party of several hundred men from the Regiment came up here to our site at one end of Nichols Field to prepare the camp, carving it out of the stinking abandoned rice paddy two days ago. We were not due to come up for two more days. But a higher authority at the behest of the moving operators moved us out. Without dinner, crowded on an inadequate number of gondola cars, we made the several-hour run to Culi-Culi, got off, waited for an hour and a half before the Regiment could be contacted and a guide sent for us, and finally arrived at the barely-completed tent city where we are to live. Supper was scrounged up at the Quartermaster: 10-in-1 rations. Only the tents are up, and the messes partially finished. Everything else is a sea of muddy grass. All five officers and 168 enlisted men are here now.

Our TAT equipment was found here, brought from the ship by a special detail.[5] Quite a number of items were lost during the careless unloading operations. Every time colored troops do something, it goes wrong.

Personnel—Pvt Lee returned for duty, and Pvts Glenn and Kepka were hospitalized.

THURSDAY—2 AUG 45

We are working on our camp only. Tentatively, plans call for our assignment on 1 Sep to the 6th Army for the invasion of Japan. We are to go in with the follow-up wave, on D plus 1, as part of a Special Engineer Brigade to build air strips.

FRIDAY—3 AUGUST 45

Pvt Glenn returned to duty. Cpl Groom was returned to Regimental Headquarters as Clerk and Pfc Mannion relieved of his Temporary Duty assignment there. The Personnel Section is displeased, preferring the latter man.

SATURDAY—4 AUG 45

The carving out of the camp continues.

Personnel—Williams was transferred to Headquarters and Service Company. Chalom was reduced to Pvt for disobedience of orders aboard the ship, effective 23 Jul. Promoted as of the 1st to Pfc again, he was one of the following list: Adams, Begale, Brigman, Boardman, Boyd, Harlan Bullock, and Kemper. Rossman was hospitalized.

SUNDAY — 5 AUG 45

Pfc James Wilson was hospitalized and dropped from the rolls. Also dropped from strength were these hospital cases: Woodward, Fisher, Tecs 5 Nance and Whitaker, and Pfcs Kepka and Rossman.

MONDAY — 6 AUG 45

There is still no equipment; consequently, there is no work other than housekeeping and camp building. Lumber is being obtained to frame and floor the tents.

Pfc Woodward returned from the hospital for duty.

TUESDAY — 7 AUG 45

Pfc Woodward was dropped to the hospital again. He is being returned to the States for surgery on his bad eye which reopened aboard the ship and to which sight is returning.

WEDNESDAY — 8 AUG 45

Still no work. Can anyone be blamed for bitching about being rushed here from Europe to work only to find no work?

THURSDAY — 9 AUG 45

Lt Moseley sent out one of his squads for a guard detail to the Regimental Motor Pool located same three miles from camp. Sgt Good is in charge.

Personnel — Whitaker rejoined the Company; Mosier, Goding, and Baker were dropped from strength upon being hospitalized.

THURSDAY — 9 AUG 45 TO FRIDAY — 10 AUG 45

No change in operations. The war talk which we have been hearing almost ceased last night — the 10th — when about 2300 hours there was a radio announcement of Japanese peace terms being offered the Allies after the Potsdam Conference. Everyone started celebrating. The men drank up their beer ration. It looks as though the war may be over. Everyone is talking of going home.

SATURDAY — 11 AUG 45

Cpl Christopher was hospitalized and dropped from strength. The peace talk persists as does the talk of home.

SUNDAY — 12 AUG 45

Tec 5 Whitaker returned to the hospital, dropped from the rolls.

MONDAY — 13 AUG 45

The Regiment finally received battle participation credit for Ardennes and Central Europe. We now wear five bronze stars on our European Theater of Operations ribbons, and have this painted on all jeeps. This has caused a mild furor in Manila where MPs think a General officer rides behind the stars and therefore give him priority on the roads.

Tec 5 Kaczanowski received his Purple Heart for the wounds received at Diekirch, Luxembourg, today.

TUESDAY — 14 AUG 45

Tec 5 Maas was hospitalized and dropped from strength. Sgt Good returned with his squad from the guard detail.

WEDNESDAY — 15 AUG 45

The end of the war with Japan was announced this morning. Even though V-J Day will not be proclaimed until surrender terms are signed, everyone feels elated that there are no more wars in the world, nearly fourteen years after the fighting started in Manchuria. We are slated for Japan, though.

Tec 5 Nance was reassigned to and rejoined the Company.

THURSDAY — 16 AUG 45

Pfc Wright was hospitalized and dropped from the rolls.

FRIDAY — 17 AUG 45

I guess this was a big day on Nichols Field. Lord Louis Mountbatten arrived here this morning, and the Chinese delegation is rumored due at noon. General MacArthur's plane was warming up. There are British and Australian officers to be seen in town, too.

It seems safe to think about it now. On Wednesday last the Colonel announced that the Regiment was destined to go to Tokyo: this, the day of war's end. The Regiment almost blew up. Lots of wild talk filled the air: civil rights and citizenship and trash like that, instead of soldiering. A goodly number of misguided civilians signed a petition without meaning which, fortunately for them, was squelched before the signatories became eligible for a General Court-Martial. And yesterday there were more meetings within the unit than ever before. The Colonel called the First Sergeants and then the noncommissioned officers of the first three grades. Officers' meetings were held. Even the Censors met. Granting all disappointments — and there are keen ones — it seems not only wrong but silly to bitch after not quite 18 months of overseas service when there are men here with 36 months who are not behaving like children.[6]

Personnel — Rajewski, Jessie Bullock, and Bailey entered hospital and were dropped from rolls. Rossman was reassigned to and joined the Company.

SATURDAY — 18 AUG 45

POM again, only this time there is not much of it as we have barely anything to get ready for loading.

Personnel — Goding and Kruk were reassigned to and rejoined the Company today.

SUNDAY — 19 AUG 45 TO WEDNESDAY — 22 AUG 45

Levey and Glenn were hospitalized and dropped from strength. Cpl Christopher and Kepka were reassigned to and joined the Company Tuesday.

THURSDAY — 23 AUG 45

Rajewski and James Wilson were reassigned to and joined the Company today.

SATUDAY — 25 AUG 45

Mosier and Glenn were reassigned to and joined the Company. The following men were transferred out for shipment to the States: Sgt Valentine, Tecs 4 Sucher and Stuck, Cpls Hertzog and Korkin, and Pfcs Merritt and Olson.

SUNDAY — 26 AUG 45

The following reductions were made today: Cpl to Pvt — Groom and Hetler, Tec 5 to Pvt — Hart. Promoted were: To Tec 4 — Richardson and Storm, to Cpl — Francis, Mannion, and Varner, to Tec 5 — Hammack, Isham Wilson, and Hetler. Cpl Mannion replaced Pvt Groom as Company Clerk and went to Regimental Headquarters on Temporary Duty.

MONDAY — 27 AUG 45

Between 1200 and 1500 today we embarked on LST 798 and for all practical purposes departed the Philippines. Each Company was assigned an LST for the crossing, and we are on the Flagship of the flotilla.

Personnel — Isham Wilson was promoted to Cpl and James Wilson to Tec 5.

AT SEA

ABOARD USS LST [LANDING SHIP TANK] 580

EMBARKED 1600 28 AUG 45 — DEBARKED 1615 14 SEP 45.

TUESDAY — 28 AUG 45

We are off again! This morning we were ordered off LST 798, told to load our motor vehicles on 908 and ourselves on LST 580 which means traveling with Headquarters and Service Company and "City Hall," something not considered a privilege.[7] Most of our men are sleeping on the open deck. The original assignment would have made this the most pleasant crossing of the four. It must have been too good to last.

WEDNESDAY — 29 AUG 45

At 0830 this morning we got off the beach and anchored in Manila Bay to await sailing orders.

THURSDAY — 30 AUG 45 TO FRIDAY — 31 AUG 45

These two days we spent sitting blithely swinging at anchor in Manila Bay.

SATURDAY — 1 SEP 45

These promotions were made today (Pvt to Pfc): Hart, See, Franklin Smith, Soares, and Wojewodski. We continue swinging at anchor, blithely.

SUNDAY — 2 SEP 45

V-J Day was proclaimed directly following the signing of surrender terms. We heard it on the radio between 1030 and 1110 this morning.

MONDAY — 3 SEP 45

Sailed today about 1400 hours, after sitting at anchor for an even week. There are 32 LSTs in convoy together with one DE [Destroyer Escort, two PC's [Patrol Craft], and an LCI [Landing Craft, Infantry].

TUESDAY — 4 SEP 45 TO THURSDAY — 6 SEP 45

The ship, considerably smaller than any of our previous craft, pitches and rolls a goodly bit, but the small amount of seasickness indicates a hardy breed of sailor-soldiers.

FRIDAY — 7 SEP 45

Many of us saw our first burial at sea today. A soldier died on LST 504. The ship pulled out of convoy for perhaps ten minutes and stopped. The flags of the flotilla were lowered, there was a splash as the weighted body went into the water, the flags were half masted for a moment, then raised. The LST started forward again, rejoined the convoy, and everything was over.

SATURDAY — 8 SEP 45 TO TUESDAY — 11 SEP 45

The weather continues cloudy and rainy by and large, although there were several good days full of sunshine.

WEDNESDAY — 12 SEP 45

Sighted Honshu Island about 0800, and anchored in Yokohama Bay about 1310 this afternoon.

THURSDAY — 13 SEP 45

Beached at 1000 this morning but instead of unloading, were ordered to proceed to Tokyo. Departed about 1300 and arrived in Tokyo Bay about 1530, where we anchored again for the night.

FRIDAY — 14 SEP 45

In the morning we were ordered back to Yokohama for unloading. Finally, we beached about 1530 and started unloading at 1615 this afternoon. The last work detail completed unloading rations at 0200. The Company spent the night on the beach.

JAPAN

TOKYO, JAPAN

ARRIVED 1100 15 SEP 45—DEPARTED

JAPAN GREETED THE FIRST American troops ashore, not with spears or grenades, but with silence. Fearful of what would happen, they were almost supine in public. Within weeks, however, this began to shift to an acceptance of the presence of *Gaijin* on imperial soil. Japanese attitudes and fears formed the unrealized — to Americans — framework for the occupation. U.S. guards took post around the Imperial Palace, and Gen. Douglas MacArthur moved into the Dai Ichi building, both signs of the changed fortunes of the warring parties.

Immediately after the surrender, the Japanese people suffered from the psychological devastation of defeat, malnutrition, and homelessness. Within the context of government propaganda exhorting them to fight and die "like shattered jewels," along with the indoctrination of the "Code Bushido" and its corollaries, the people were forced to endure the desecration of "the imperial soil . . . by well-fed, superbly equipped, supremely confident GIs." Destruction of their merchant marine and damage to their rail network combined with insufficient crops in 1944 and even worse crop failure in 1945 to leave most Japanese on below-subsistence ration levels. In most major cities the U.S. bombing campaign had left over half the dwelling units uninhabitable or destroyed. In Tokyo 65 percent of the residences were destroyed. Nationwide, 43 million people were homeless. The rich neighborhoods seemed to be less damaged than the poor ones, leaving most returning veterans and civilians without homes to come back to and families who were starving.[8]

High officials in the military and in the government turned to embezzling supplies and equipment, and a black market (especially for food) developed. The lack of a "tradition of responsibility towards strangers or of . . . philanthropy, or of tolerance or even genuine sympathy" for the unfortunate "improper" people like war widows, homeless veterans, the disabled, and the handicapped exacerbated the misery. Inflation and crime increased everywhere. The government created the Recreation and Amusement Association, recruiting some women to serve as prostitutes in "recreation and amusement" centers to protect the virtue of most others. The pay was a great incentive to set honor aside. Of course, venereal-disease rates quickly rose, and some 90 percent of the women thus employed tested positive, as did many soldier-clients. In one unit 70 percent of the men had syphilis, 50 percent gonorrhea.[9]

SATURDAY — 15 SEP 45

The first half of the company arrived at billet area in Tokyo, Japan, at 1100 and returned trucks to beach for the remainder of the Company who arrived about 1700 hours.

Present were: 5 officers and 157 enlisted men. The Company Clerk, present as the 158th man, was with Regimental Headquarters.

SUNDAY — 16 SEP 45 TO MONDAY — 24 SEP 45

The government provided a number of Japanese laborers to work as dirty men or Kitchen Police, and to do the sweeping and mopping in the building we occupied, and this helped somewhat in setting up housekeeping. Although at times we all wondered if it was worth the effort that had to be made to talk to them. If it had not been for Pfc Biarruel, Company "B's" hired help would never have received any instructions except by sign language. This linguist ended up by being able to talk to the natives in every country we entered, however sketchily.

The beginning of our stay in Japan was similar to what had happened in

Manila—nothing. On the 18th Pfc VanLuyn was hospitalized. On the 20th, while reconciling a difference of one in the enlisted strength of the Company, we found that Pfc Kruk had not been dropped from the rolls when he went to the hospital on the 8th of August but that he was picked up upon his return later, and this error was corrected.

On the 21st the first transfer was made when Pfc Peterson went over to Headquarters and Service Company to work in the S-4 section. What with the high pointers being earmarked for shipment home, the line companies were told they would furnish the personnel to keep Headquarters and Service Company up to T/O strength with qualified men. Nobody was especially pleased with the prospect of transferring out the good men, but it was a necessity. On the same day Pfc Michek was hospitalized with an injured ankle.

Pfc Erickson, with an ASR score of 77, was really sweating out news of shipments home. Since his injury in an air attack in Germany on April 17th, he had been waiting to be awarded the Purple Heart—and what he was waiting for was not the decoration, but the five points that came with the medal. Finally, on the 23d, the award was made by the Regimental Commander. His score jumped to 82, and he was eligible to go home.

TUESDAY—25 SEP 45

And go home he [Erickson] did on the 25th, when he was one of 25 enlisted men relieved from assignment to the 1303d, transferred in grade to the 103d Infantry Regiment of the 43d Division, and prepared for shipment to the States. This first batch of 80 pointers, was comprised of the following men: Sergeants Bennett, Boyes, Chavez, Faucett, and Shaffer; Technician fourth grade Richardson, Corporals Anselmo, Cooper, Mosley, Rajewski, Runions, and Isham Wilson; Technician fifth grade Goding; Pfcs Cagle, Cederberg, Clary, Erickson, Gossett, Hart, Kruk, Chester White, Houston White, and Zins; and Privates Groom and Platter. It was neither easy nor pleasant for the Company to lose these men who had been an important part of the organization, most of them, from nearly its date of inception. But they had done their duty and were leaving.

The same day orders were received relieving Pfc VanLuyn from assignment, and he was dropped from the rolls to the 43d Field Hospital. And Lt Reese was relieved of assignment reassigned as Regimental Special Service Officer, to understudy high-point "Mike" Neeson. Company "B" was back again to four officers. If any one thing was our jinx, that was it.

WEDNESDAY — 26 SEP 45

More men were transferred to Headquarters and Service Company today: Tec 5 Catlett, and Pfcs Jiminez, Kolbusz, and Murray. Pfc Fout was placed on Temporary Duty with Regimental Headquarters as Company Clerk, relieving Cpl Mannion who began preparing to go home with the next batch, soon going, the 70 pointers.

Captain Hall departed for three days of leave on Okinawa, and Lt Deveikis assumed additional duty as acting Company Commander. Before his departure, the Captain outlined the shape of things to come as he learned it at a Company Commanders meeting. If what he said were to come true, then before he returned the Regiment would be unrecognizable for the departure of everybody was contemplated.

THURSDAY — 27 SEP 45

On the 27th sixty enlisted men were relieved of assignment, transferred to the 103d Infantry, and departed in the evening to go directly to the ship that would take them home, the USS "Hershey." This happy contingent was headed by 1st Sgt Douds who took along the very heart of Company "B": S Sgts Cooke, Dunbar, Jenkins, Murphy, and Sheridan; Sergeants Cunningham, Good, Hawke, Hasenkamp, Pennington, and Shouse; Corporals Jasperson, Mannion, and Varner; Technicians fifth grade Bordonaro, Hammack, Clark, Ekenberg, Giambrone, Kaczanowski, McGlashan, Mosier, Nance, Roush, and James Wilson; Privates first class Begale, Bersaw, Biarruel, Blagun, Brehl, Chalom, Dobrin, Freeman, Gainey, Gavinskv, Jacobs, Willis Kellevr, James Kelly, Kemper, Kustera, Michek, Minkemann, Mitchell, Ormsby, Pisani, Pyke, Reed, Ripley, Rosinski, Schara, See, Silvers, Charles Smith, Westbrook, Westphal, Willmeth, and Wojewodski; and Private Mesner.

In order to make it, Pfc Michek was removed from the hospital where he was getting around on crutches. In the Company Tec 4 Storm improvised a practical bonvoyage present — a pair of crutches: so riding in the ambulance, even the sick made it to the ship where they embarked at nine o'clock.

The Company was left with 67 men and four officers. For all practical purposes, it was finished. The Regiment was finished. There remained but the removal of the remaining stragglers and deactivation.

THE REGIMENT BECAME the construction unit for the Tokyo general headquarters. The men were quartered in the Olympic Village, built for the 1940 Olympics, which never took place. Among the facilities were a huge swimming pool and the Mitsubishi Auditorium, where a festive rodeo would be held — which must have been strange behavior indeed by Japanese standards. They were assigned to Army Service Command C, 5202nd Engineering Construction Brigade for administration.[10]

FRIDAY — 28 *SEP* 45

Today Lt Moseley did the second job assigned us since our arrival in Tokyo. He with a small detail checked with the Engineer's Office at Eighth Army Headquarters about eight heaters. From there he was referred to USAS-COMC [U.S. Service Command C] where he learned the ship bearing the heaters was not unloaded yet, and besides somebody in General Headquarters wanted nearly more than there were on board. So the requisition was left and notification for pickup and delivery would come when the heaters were ready.

The first job, construction of an RCVR [radio receiver] building for General Headquarters near the Tokyo beach, started with the grading of the site. It lasted for one day. It also ended with the grading, because, that finished and reported last Saturday, the 22d, nothing further was assigned.

Of the four officers, one is inoperative for the Company. Assigned to Engineer Liaison at General Headquarters, Lt Larew spends his days inventorying buildings being requisitioned for the expanding OCCUPATION ARMY. We call it counting windows in Tokyo.

During the first days here the Headquarters LST (no. 171) pulled in and Lts Moseley, Larew, and Reese with their platoons working around the clock unloaded the 400 tons of cargo in 18 hours, transporting it to various places in Yokohama and Tokyo.

Personnel — VanLuyn returned from the hospital, Sharpe and Howard Smith were transferred to Headquarters and Service Company.

SATURDAY — 29 *SEP* 45

When every Company Commander started recommending his remaining men for the promotions to positions left vacant by the mass removal of the heart of the Regiment, the powers that be stepped in with a modified T/O which excluded these positions: S Sgt Construction Foreman, Sgt Railroad Construction Foreman, Tec 4 Carpenter General, Tecs 5 Carpenter Bridge, Welder, Blacksmith. In addition, except for Companies "B" and "E," running the battalion messes, no kitchen personnel promotions were authorized. Following this directive we promoted the following men to their T/O positions: 1st Sgt Boling; SSgts Fritz, Lanzisera, and Russell; Sgts Christopher, Kuhn, Oden, Francis; Tecs 4 Turner and Schoenfeld; Cpls

Plucinnik, Fout, Stagner, and Zizzi; Tecs 5 James and Maurice Jones, Levey, McDaniel, Livera, and George Miller. Adams and Kepka were transferred to Headquarters and Service Company.

SUNDAY — 30 SEP 45

Captain Hall returned from leave, resumed command of the Company.

MONDAY — 1 OCT 45 TO TUESDAY — 2 OCT 45

Glenn and Puccio were promoted to privates first class, on the 1st.

WEDNESDAY — 3 OCT 45

Hetler and Mark Keller departed for return to the States. Prigman was transferred to Headquarters and Service Company. Promoted were: Reimer to Tec 4 (replacing Lewanski who departed for the States), Bruehler and Wania to Tec 5.

. . . .

FRIDAY — 5 OCT 45

Lt Reese relieved of assignment and reassigned to Regimental Headquarters, primary duty Special Service Officer. Tec 5 Fohlio was hospitalized.

SATURDAY — 6 OCT 45 TO SUNDAY 7 OCT 45

Pfc Puccio was transferred to Headquarters and Service Company, 6 Oct.

MONDAY — 8 OCT 45

Lt Deveikis was ordered to start his return to the States, and departed. Capt Hall was relieved of assignment and reassigned as 1st Battalion Commander. Lt Moseley assumed command of Company "B."

CAPTAIN HALL WAS TRANSFERRED under Special Orders No. 158, Headquarters, 1303rd EGSR, dated 8 October 1945, from Company B to 1st Battalion headquarters as commander; Paragraph 6 designated the captain as a "Summary Court." Second Lt. Richard Reese was transferred from Company B to regimental headquarters as special services officer.[11]

. . . .

THURSDAY — 11 OCT 45

Tec 5 Roberson was transferred to Headquarters and Service Company. Pfc Boyd was hospitalized.

Major Gonzalez, Japan, 1945. Courtesy Dwight Gowdey

FRIDAY — 12 OCT 45

Our first job since the European Theater of Operations was started today. A radio receiver building containing some four thousand square feet of floor space was started on one of the artificial islands in Tokyo Bay. The job was let to Jap contractors, who are to do actual construction. The company is responsible for supervision and procuring and hauling materials.

The Japanese survey party laid out the building and set grade stakes. One squad went out to remove some logs from the shoulder of the road at the site to give trucks access to the area when fill hauling starts.

SATURDAY — 13 OCT 45

Today the Company received its first regular pay since we left the European Theater of Operations. The old Eagle screamed four months worth, but he's still behind.

Company B officers, Japan, 1945: Captain Hall, Lieutenant Larew, Lieutenant Deveikis, and Lieutenant Mosely. Courtesy Dwight Gowdey

Receiver Station — 72 sections 6" × 12" diameter concrete pipe hauled for sub-floor cable conduit. 270 board feet of one inch lumber drawn from USASCOMC Engineer Dump for forming floor slab and 6 motor generator foundations.

SUNDAY — 14 OCT 45

Receiver Station — 180 board ft 3" × 6" form material hauled and generator foundations locations surveyed.

MONDAY — 15 OCT 45

Receiver Station — Began hauling gravel for 2" fill under building using 30 Jap laborers to load trucks. Generator foundations excavation started.

TUESDAY — 16 OCT 45

Receiver Station — Japs began forming generator foundations. Gravel hauling continues.

. . . .

THURSDAY — 18 OCT 45

Receiver Station — forms completed. Hauling continues.

FRIDAY — 19 OCT 45

Receiver Station — Started hauling gravel at night using "E" & "F" Company trucks as well as our own.

SATURDAY — 20 OCT 45

Lt Ingle (Regimental Adjutant) hit a lick when he tracked down authority by which the ASR for men overseas raised 8 points instead of points based on recomputation as of 12 May. Two officers and 50 enlisted men got from 1 to 2 more points!

Receiver Station — Building forms for conduit terminals. Hauling of fill continues, in addition to hauling sand and gravel for concrete.

SUNDAY — 21 OCT 45

Tec 5 Foglio returned from hospital and returned to duty in the kitchen. S Sgt Russell had been under a burden to feed the whole Battalion with a short handed kitchen crew.

Receiver Station — No work — Sunday

MONDAY — 22 OCT 45

No work today because of rain. The Japs eat so little they cannot survive if exposed to inclimate [*sic*] weather.

Personnel — Pfc Boyd from sick in Hospital to duty.

TUESDAY — 23 OCT 45

Receiver Station — Hauling of gravel continues day and night. A Jap type concrete mixer was set up and water piped in.

Personnel — Pfc Leo Killer was promoted to Tec 5 and 35 replacements joined the Company. They were all fresh from the States with 5 months in the Army and only Infantry Basic behind them.

. . . .

THURSDAY — 25 OCT 45

Receiver Station — Began hauling conduit, hauling gravel at nights using 2d Battalion trucks, and setting up Japanese concrete mixer.

FRIDAY — 26 OCT 45

Receiver Building — Poured one motor foundation, began building forms for conduit. Hauling gravel for fill continues.

Personnel — Today we made non coms for the new 2d Platoon, which we received on the 23d. Sgt Oden became S Sgt; Cpl Stagner, Tecs 5 Lange and Levey became his sergeant squad leaders. In the 1st platoon, boosted up to replace men moved to the 2d: Pfc Thornton made Sgt, Tec 5 Wania made Cpl and Pfcs Bogue, Farr, and VanLuyn became Tecs 5. Tec 5 Roberson returned home to "B" Company from "H&S" Company and Pvt Schoolcraft was transferred to Headquarters and Service Company.

SATURDAY — 27 OCT 45

Receiver Building — Completed fill last night. Now hauling gravel to mix concrete; and lumber for building. Bottom half of 6 motor foundations poured. Forms being set for top halves. Began laying conduit. Job 25% complete.

Personnel — Pfc Perrine transferred to Headquarters and Service Company.

SUNDAY — 28 OCT 45

Receiver building — All foundations poured. M/Sgt Helm, one time 1st Battalion Sergeant Major went home.

MONDAY — 29 OCT 45

Work on Receiver building continues. Again hauling sand for concrete.

Personnel — Cpl Plucinnik from duty to sick in Hospital. 71st Evacuation Hospital. [He was "returned to duty from the hospital" on 2 November 1945.]

TUESDAY — 30 OCT 45

Receiver Building — For reasons unknown, the Japs did not show up for work today, after a poor showing yesterday. Our trucks continue to haul lumber.

32 men joined the Company today but as yet are not assigned.

WEDNESDAY — 31 OCT 45

Everyone from Office of the Chief Engineer down to and including our Battalion Commander is on the war path. Capt Hall threatens to jail some Jap contractor if they don't go to work. Company "B" continues to haul lumber and cement.

THURSDAY — 1 NOV 45

5 Jap workers finally showed up today and resumed work on receiver building.

Personnel — Promoted to Pfc were: Pvts Routh, Schlosser, Hastings, Khoury, Waldeck, Stenback, Stone, Tessier, Luther Smith, Shirel, Cooper, Simpson, Simkin, Stamper, and Rohder.

FRIDAY — 2 NOV 45

Work on receiver building continues.

Personnel — the 32 men who joined the Company on 30 Oct were assigned today. . . . Pfc Simpson transferred to Headquarters and Service Company.

SATURDAY — 3 NOV 45

Receiver Building — Began prefabricating building.

Personnel — "B" Company has a 3d Platoon again. 41 enlisted men assigned and joined the Company today. . . . With these men we made the 3d Platoon and some promotions. Sgt Francis became S Sgt, Tec 5 Mataitis became Sgt, and with Lancaster and Nix are S Sgt Francis's squad leaders. Pfc Schlosser and Pfc Stone were promoted to Sgt and Cpl respectively in Headquarters Platoon to work for Lt Larew at the Office of the Chief Engineer. Pvt Hardin and Pvt Alvarado also became Corporals. Pvt Ketcherside transferred to Headquarters and Service Company.

SUNDAY — 4 *NOV* 45

Continue work on receiver building.

Personnel — Oh happy day — Promotions are in order. Lt Larew made 1st after the rest of the officers in the regiment up for promotion were turned down except Lt Davis who made it a week ago. Relieved from assignment and transferred to Headquarters Battery, Division Artillery, to go home with the Americal Division are S Sgt Russell; Sgt Kuhn; Tec 4 Reimer; Cpls Plucinnik and Zizzi; Tec 5 Kelly Joseph; and Pfcs Gamble, Hughes, and Paffie.

MONDAY — 5 *NOV* 45

Japs continue working on receiver bldg.

Personnel — Oh unhappy day. The regiment has become a repple depple [replacement depot]. We are getting men because there is no place else to feed and clothe them, and the strain in "B" Co's battalion mess hall is terrific. 9 men assigned and joined the Company today. And we expect more tomorrow. . . .

TUESDAY — 6 *NOV* 45

Receiver building — Work continues. Job 30% done.

Lt Larew continues to work for the Office of the Chief Engineer which requires all his time, though he expects to have to take over the Company on the 15th.

Personnel — Today 16 men joined the Company. . . . Company strength 2 officers and 177 enlisted men.

Of these 177 enlisted men, 98 were new. Of the 168 who came to Japan, only 79 remained.

WEDNESDAY — 7 *NOV* 45

Work continues on Receiver building.

Today "B" Company undertook the task of establishing a mess hall for Company "A" with personnel we trained for "A" Co. Pfc Waldeck of "B" Company will be transferred to "A" Company for Mess Sergeant. Company "C" will be the next to open a kitchen.

Personnel — Enlistment drive started today for the Regular Army. All units in 8th Army will have 20% of their personnel enlist or the unit commander will reply by endorsement why not.

Pfc Bullock honorably discharged from the Army of the United States for the purpose of enlistment in the Regular Army.

THURSDAY — 8 NOV 45

Receiver building — Motor foundations and conduit are finished. Reinforcing rod and forms are being placed to pour floor slab. Job 1/3 done.

Personnel — Pfc Bullock — enlistment Regular Army — attached unassigned.

FRIDAY — 9 NOV 45

Major Gonzalez, one time Company Commander, and 5 more high point officers left the Regiment for home.

Personnel — 5 more men . . . assigned and joined the Company today. Tec 5 Bogue from duty to sick in 71st Evacuation Hospital.

SATURDAY — 10 NOV 45

Receiver building — Pouring concrete floor slab.

SUNDAY — 11 NOV 45

Today was declared a holiday by General Headquarters. This afternoon the International Allied Horses Rodeo was put on by ASCOM-C in Meiji Stadium over at RSO. 55,000 people were present. A 30 minute air show by the 5th Air Force preceded the rodeo which was produced by Lt Dick Ryan of 1st Cavalry Division. (Lets give Lt Ryan a big hand.) "C" Company built all the shutes and pens for the rodeo as well as hauled stock to the grounds.[12]

MONDAY — 12 NOV 45

Continue pouring concrete at receiver building. Job 35% done.

Personnel — Pfc Waldeck transferred to "A" Co. Pvt Kassel transferred to the Medical Detachment and Pvt Bryant and Focht honorably discharged for the purpose of enlistment in the Regular Army.

TUESDAY — 13 NOV 45

Construction on receiver building continues.

Pvts Bryant and Focht enlisted Regular Army. Attached unassigned to this company. Pvt Jordan from duty to sick in 251st Station Hospital. Pvt Callaway transferred to Headquarters and Service Company.

END OF DIARY

15 NOV 45

Capt Earl E Hall O 520528 ASR 75 transferred with 7 other officers to Disposition Center, 4th Replacement Depot by Letter Orders, HQ USA Service Command C, APO 404 of 7 Nov 45.[13]

APPENDIX ONE

Types of Units

Soldiers doubtless find the names of various units precise and clear; civilians are confused enough by "company" or "regiment." What was a "Light Equipment Company" and how did that differ from an "Engineer Special Service Regiment," and what does it mean that the 1303rd was an "Engineer *General* Service Regiment"?

The U.S. Army during World War II was organized into units that were nominally consistent across the service. The basic units were formed of soldiers from a single arm (for example, infantry or artillery) or service (Transport Corps, Military Police, and such): an engineer company had all engineers, an infantry company, infantrymen. The basic element was the squad, up to sixteen men commanded by a corporal or sergeant. In this diary Company B was organized into three four-squad platoons, each commanded by a second or first lieutenant. Companies, including B, were commanded by a first lieutenant or captain. The "Table of Organization" in effect for the 1303rd called for a company to have five officers: a commander; an executive officer, who supervised the headquarters section; and three lieutenants commanding the three platoons. The company was authorized a particular number of enlisted men and officers (called "strength").[1]

A squad is the basic unit capable of performing branch functions, in this case, of performing small engineer duties such as road maintenance. A platoon was the normal unit for ordinary engineer operations such as building or repairing a bridge or removing a Bailey Bridge. A large project, such as bridging the Moselle River, might take a company. Since no basic outfit could have all the assets needed to do every possible job, specialist units, such as ponton companies, light-equipment companies (cranes), and dump-truck companies, were organized to handle the more unique and unusual equipment.

Specialist companies were typically broken up into its smaller components and attached to other commands. The 764th Engineer Dump Truck Company, for example, was attached to the 1303rd, with one platoon attached to each battalion. One squad of each DT platoon was attached to each company of the regiment for operational work, rations, and quarters. Since this DT company was a black unit, and the 1303rd EGSR a white unit, in most companies the DT soldiers were kept segregated in eating and sleeping quarters. In Company B they were included without regard to the niceties of racial separation.

Companies A, B, and C were part of the 1st Battalion, 1303rd EGSR; Companies D, E, and F formed the 2nd Battalion. Each battalion, commanded by a major or lieutenant colonel, had a headquarters section made up of men drawn from the component companies and an attached medical section. A battalion is the smallest command unit that has intelligence, supply, operations/training, and personnel function specialists.

The six "line," or operating, companies in the two battalions, together with the "Headquarters and Service" Company and medical detachment, formed the engineer general-service regiment, commanded by a colonel. The regiment performed a number of functions as it supports and coordinates its companies and battalions in accomplishing the various missions assigned to it, such as engineering design and drafting functions (as with the 1303rd).

Within the context of similar unit structures, there were significant differences between the various combat and support engineer units. The 1303rd was, as an EGSR, a support unit, not a combat unit. As such, it was assigned to the Third Army as an integral unit under army control. Third Army headquarters attached the 1303rd to the XII Corps to assist the Corps Combat Engineer Group. As a result, the Thirteen Third spent most of its European service in the combat zone rather than the communications zone.

Types of Engineer Units

The 1303rd was an "Engineer General Service Regiment." Other kinds of engineer units are mentioned in the diary. The main distinction was between combat and noncombat units. All engineer units came under the authority of the chief engineer, who headed the Corps of Engineers. Operationally, combat-engineer units came under control of the Army Ground Forces (AGF) commander, while noncombat units were under the Army Service Forces (ASF) commander.

Combat elements—the combat engineers—were at first organized in regiments. Separate companies provided pontons for floating bridging, Bailey bridges and Treadway bridges for temporary crossings, troops for mine emplacement and clearing, and assault boats for crossing rivers. Early in 1942 the AGF, commanding all combat elements, shifted from the regimental to the battalion form of organization. The result for combat engineers was to essentially divide each of their regiments into its two constituent battalions, and formerly integral capabilities were made separate companies. Battalions were combined into combat groups (equivalent to brigades).[2]

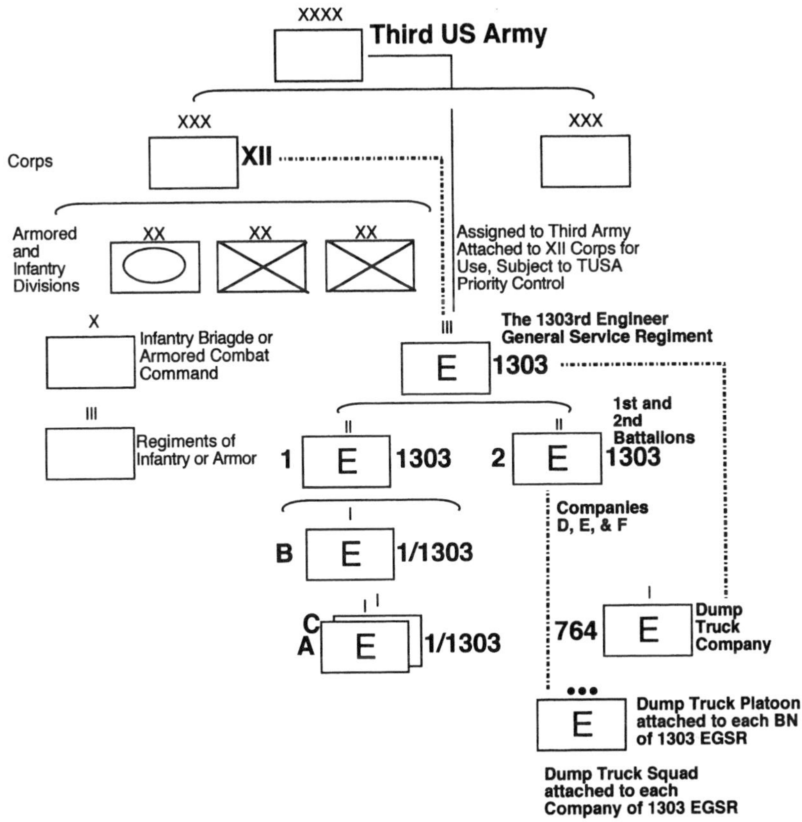

Organizational structure of Third Army, showing the relationship of the 1303rd EGSR to army headquarters, the XII Corps, and other units. Courtesy JCF

Officially, the "function [of] front-line engineers [was] . . . to clear and construct obstacles, lay [and remove] mine fields, ferry troops in river crossings, build bridges and . . . act as infantry." Those in the rear were more concerned with "building shelters, roads, ports, or air fields, and with performing supply functions." Support units, like general service regiments, were not to perform the "logistical task of construction" while under attack. They were to build and maintain roads, bridges, hospitals, prisoner-of-war enclosures, pipelines, headquarters buildings, and such.[3]

The ASF command had resisted AGF demands that general-service regiments be restructured essentially as labor service battalions but did reduce the size of the EGSRs by removing some truck and equipment elements. By mid-1943 these regiments consisted of six "line" companies (A–F) organized into two battalions and the Headquarters and Service Company. In the process of reorganization and shrinkage, each regiment also lost significant equipment stocks and the authorization for higher grades of enlisted men necessary to complete its missions; European-theater GSRs discovered that these reductions caused particular problems in operations. Of the seventy-nine GSRs raised, fifty were black units and twenty-nine were white. Many of the GSRs, particularly the black units, were regarded as mere pools of unskilled labor, which might account for the willingness of the Office of the Chief Engineer and the ASF command to accept the reductions in equipment and authorized grades.[4]

Several supporting units were created to utilize this equipment. Light-equipment companies handled the transport and operation of construction machinery. Dump-truck companies (25 white, 112 black) provided much of the missing "lift" since a GSR did not have enough trucks to move all its own men and equipment at one time. Indeed, even with a platoon attached to each battalion or a squad for each EGSR company, there remained insufficient transport available. Consequently, movement was never a single process but rather lumpy and discrete.[5]

APPENDIX TWO

Ranks and Command

Soldiers are classified as enlisted men, warrant officers, and commissioned officers. Enlisted personnel are either privates and privates first class or noncommissioned officers (corporals and sergeants). Warrant officers are specialists who have a "warrant" from the secretary of war rather than a commission; while they supervise technical operations (in the 1303rd, the sole warrant officer supervised the motor pool), they technically do not have command functions. Command is reserved to "commissioned officers," who receive their "commission" from the president of the United States, technically with the advice and consent of the U.S. Senate. Commissioned officers exercise command authority over troops.

Noncommissioned officers (NCOs) also have command authority. In the field an NCO actually supervises enlisted personnel (privates), who accomplish the task. Officers issue the relevant orders that designate *what* is to be done, *when*, *where*, and by *what unit*. At each lower level the assignment becomes more defined, and when orders finally reach the company level, a sergeant is given charge of the mission. The sergeant decides, based on his orders, which squad(s) will do particular things in particular places at specified times. His subordinate squad leaders (corporals or sergeants) will assign particular tasks to individual soldiers so that the mission is accomplished with minimum confusion and error.

The order to lay a "barrier zone" in the regiment's "area of responsibility" (AOR) comes from corps or army headquarters. The regimental commander passes the orders to his battalion commanders, who in turn pass them on to one or all of their company commanders. The company commander may designate one platoon to complete the mission or may issue orders to each platoon for their AOR. The platoon commander (a lieutenant) will tell his platoon sergeant to implement the orders. The officer might go with the platoon sergeant to lay out the barrier, or the plan might come from the regiment or battalion commander already drafted and subject to on-the-spot modification. Regardless, the platoon sergeant will call in his squad leaders and give them the orders to take certain mines out of stock, use certain soldiers in certain vehicles, and lay the mines in particular places by a designated time in order to complete the barrier zone.

Enlisted personnel wear their grade insignia on their sleeves. Privates (Pvt.) have no insignia. A private first class (Pfc.) wears a single chevron, point up. A corporal (Cpl.), grade 5 (army numbering descends in order—from 7 to 1—as grade and pay increase), wears two stripes or chevrons point up. A sergeant (Sgt., grade 4) wears three chevrons. A staff sergeant (S.Sgt., grade 3) wears the same three chevrons, point up, with a arced "rocker" curving from the outer corner of the bottom edge of the lowest stripe to the other outer corner. A technical sergeant (T.Sgt., grade 2) adds a second "rocker." In grade 1 are both master sergeants (M.Sgt.), wearing three chevrons and three rockers, and first sergeants (1st Sgt.), with three chevrons, three rockers, and a diamond-shaped lozenge in the space between the first stripe and first rocker. Some troops are considered "specialists," and even though they do not necessarily command other soldiers, they deserve higher pay for their greater skill and hence require a higher grade. These men are classified as "technicians." A Technician grade 5 (Tech 5) wears corporal stripes with a "T" underneath. Similarly, a Technician grade 4 (Tech 4) wears sergeant stripes, again with a "T" underneath. A Technician grade 3 (Tech 3) wears staff sergeant stripes with a "T" between the bottom stripe and the rocker.

A company "first sergeant" was not a grade 1 but probably grade 2 or grade 3 sergeant. Similarly, a battalion "sergeant major" was usually a grade 3 staff sergeant, perhaps a grade 2 technical sergeant. The regimental "sergeant major" was a grade 1 first sergeant or master sergeant. In the post–Vietnam War era, enlisted grades were rearranged, with the army introducing the grade of sergeant major, which entails the functional position as well as the title.

Warrant officers come in two classes, warrant officer and chief warrant officer. The insignia of a warrant officer is a gold bar covered with two squares of brown enamel, while the chief wears a gold bar with a brown enamel border to a inner gold rectangle. These officers are entitled to the courtesies and respect accorded commissioned officers (including salutes) from enlisted personnel, are called "Mister," and receive the pay of a second or first lieutenant, respectively.

Commissioned officers are classified as either company- or field-grade or general officers. The lowest in rank is the second lieutenant (2nd Lt.), who wears a gold bar, hence the common appellation "butter bar," They are often called "Shave-tails" and generally regarded as officers in training. After successfully serving for twelve to eighteen months as commander of a platoon, a second lieutenant (officer grade 1 or O-1) typically gains automatic promotion to first lieutenant (1st Lt.). This brings a silver bar, a pay raise, more responsibility, and more confidence in his abilities from the men under his command. First lieutenants (O-2) sometimes command companies and may serve as company executive officers or battalion staff officers. Normally, a captain (O-3) commands a company. Captains wear two joined silver bars, called "railroad tracks" because of their appearance. Their superiors often envy captains since they actually command troops directly. Captains are out in the mud with the men, while majors are either in a staff tent or a headquarters building.

Majors (O-4) and lieutenant colonels (O-5) wear oak leaves (gold and silver, respectively). In general, majors are staff officers, while lieutenant colonels command battalions or even regiments; in the 1303rd majors commanded battalions. A colonel (O-6) wears a silver eagle. They often command brigades or serve as staff officers for higher commands (divisions, corps, armies). The only general officers noted in the diary are a lieutenant general, George S. Patton, and a major general, Manton P. Eddy (commanding Third Army and XII Corps, respectively).[1]

APPENDIX THREE

Training the Soldier

A new recruit for any of the services began his training at a reception center, where he got a range of mental, physical, and aptitude tests and interviews; inoculations; a haircut; and an issue of basic equipment. Army enlistees were then sent to a replacement depot, where they received additional equipment and were assigned to a barracks and effectively quarantined for two weeks. During this early period, the men were tested and interviewed (hobbies — especially amateur-radio operation — were often more valuable than job training), and assigned a military specialty and branch. From there they went to either individual training or to a unit-training center like Camp Ellis.

This period of drill, training films, more tests and inoculations, and instruction in basic military courtesies included regular runs through an obstacle course that became progressively harder as the soldier became faster and received more equipment to carry. Rifle training was central; even engineers recognized that "one thing that is more important to the soldier than anything else is to be able to shoot straight and fast." Engineer training was reduced in late 1941 to eight weeks, one week of which (15 percent) was devoted to rifle training. After 15 March 1942, the training period was expanded until it reached twelve weeks. The added time reduced stress on facilities and trainers and allowed for instruction in demolitions, obstacles, bridging, and road construction. Fort Belvoir, Virginia, and Fort Leonard Wood, Missouri, served as training centers for engineers.[1]

Corps of Engineers training facilities were expanded in 1943–44 with the addition of two engineer-unit training camps, one at Camp Claiborne, Louisiana, and the other at Camp Ellis, Illinois. Ellis was within the Sixth Service Command Area of the Army Service Forces, which controlled the post complement. Col. Robert D. Ingalls commanded the engineer group at Camp Ellis. By August 1943 experienced construction men were fewer than expected, which meant an even greater need for training. The colonel had the training units expand the camps themselves as they took in new and often inexperienced men as a way of combining practical instruction with the need for additional facilities.

Ingalls emphasized marksmanship and personally supervised concentrated eight-day periods of instruction. On average 97 percent of the recruits in his camps qualified as marksmen. The colonel also stressed the need for physical conditioning and more-realistic training, telling his unit officers (on 31 July 1943): "Put the pressure on as hard as you may, yet you still will not approach the condition of fatigue and hardship your units will soon face. . . . Every training task must be approached as though it were a battle mission."[2]

Designed for 7,400 men, Camp Ellis held 10,500 by 7 September 1943. Two EGSRs and four dump-truck companies were under canvas until October. The official history acknowledges that most of these recruits were "fillers [who] were either men reclassified from a limited duty status, or eliminated as surplus overhead" by other units. An inspector from the Office of the Chief Engineer (OCE) noticed trainees with back braces, deformed limbs, and glass eyes.

Conducting an orderly course of training was impossible since men were continually being received, examined, retained, or (more often) shipped out or discharged. By late November 1943 things began to settle down, and 3,940 of nearly 8,000 fillers were reclassified and either reassigned or discharged. At this time there were seven EGSRs at Ellis, and they needed an aggregate of 3,000 replacements, about 2,000 of whom should have been "construction foremen, construction carpenters, bridge carpenters, electricians, utility repair men, riggers, quarrymen, and blaster powder men." To remedy the deficit, specialist schools were opened to teach the needed skills, with special courses given to the officers. Ellis had one officer for every 144 men, and specialist training fell to the units, using whatever specialists they already had. In addition to this instruction, the schedule called for night exercises and a four-day march despite the large number of men only marginally capable of such rigorous activity. At Ellis this march covered 125 miles of road. In addition, in November and December 1943 the trainees built two landing-mat airstrips, repeatedly, since Ingalls was not happy with the quality or the speed of their work.[3]

Ingalls arranged the obstacle courses at Ellis so that troops moving between training areas used them en route. Not surprisingly, the system was described as "extreme," and there were numerous "gripe letters" sent to politicians, the Corps of Engineers, and the War Department. In January 1944 the 125-mile march was abandoned. Perhaps the fate of the 1317th EGSR helped doom this exercise. Conducted in "severe winter weather," they suffered 50 men absent without leave (AWOL) on 30 November, 112 missing by mid-December, and 139 AWOL at the end of December.[4]

The training at Ellis must have worked, though. The three sister units, the 1301st, 1303rd, and 1306th EGSRs, which trained together at Ellis and served together in the Third Army, were among only eleven EGSRs recommended for the Meritorious Service Unit Plaques by the OCE on 20 November 1946. The XII

Corps "reported magnificent support from the 1303rd" in its after-action report for March 1945.[5]

MTP 5–1, "Manpower Training Program" number 5–1, was the doctrine and schedule in force when the 1303rd EGSR formed in the summer of 1943. As part of this training program, the regiment built facilities at the camp as needed. The program included the schedule of time allotted to each skill, the order in which they were to be done, tactical maneuvers and conditioning physical training, and the hike. Colonel Ingalls's implementation of MTP 5–1 was within the scope of and consistent with the spirit of the program, even though the plan did not, for example, specify making men who were rifle qualified only weeks earlier requalify.[6]

APPENDIX FOUR

Equipment, Bridges, and Rations

Equipment, bridges, and rations are almost equally familiar and unknown to most readers. Some equipment is rather commonplace — the "jeep" or 1/4-ton truck, for instance. Others are possibly familiar — for example, the D-4 and D-7 tractors by Caterpillar are still made, though they have changed dramatically over the years; the R-4 is the internal-combustion-engine version of the small tractor ("D" indicates diesel power). Less obvious are draglines — the bucket on a line, "thrown out and then drug back by the crane."[1] These were attached to a crane, but which type of crane — there were Quickway and BE, each with 1/4-, 1/2-, and 1-yard capacity. Quickway was mounted on a 6×6 truck, BE on a half-track chassis.

A variety of manufacturers supplied the 6×6 trucks; apparently all those used by the 1303rd were from General Motors. A 6×6 truck had six wheels, all powered (thus, 6×6), and was nominally rated at 2 1/2 tons capacity. Dump trucks used the same chassis.

Bridges built by Company B, 1303rd EGSR, were of two types: Bailey bridges and timber-trestle bridges. After training at Camp Ellis, the unit never again assembled a ponton bridge. ("Ponton" is the correct spelling, pronounced "pon-than," not "pontoon.") The Bailey Bridge was a temporary structure, quickly assembled from prefabricated, standardized parts. Quick, strong, and durable, Bailey's were assembled on land and then launched, using a "nose" assembly, across the space. (Earl Hall said that disassembly was easy. First, the flooring was removed, then the pins holding the top piece in place were removed and a crane lifted the 600 pound panel off. The stringers were unclamped and the bottom panels came apart.) These spans were usually replaced by a timber-trestle bridge.

Rations remain the subject of complaints among soldiers. The army, through the Quartermaster Corps, has tried hard to develop the best rations possible with the technology available at any given time. During World War II, quartermasters developed the "C-ration," "D-ration," and "K-ration" to replace the packaged meals available in 1941—which were pretty poor things indeed.

The K-ration was a prepackaged daily ration that included a can of meat, a packet of powdered coffee, a cereal bar, a fruit bar, a candy bar, biscuits, gum, cigarettes and matches, water purification tablets, toilet paper, salt tables, can

Pile-bent bridge, Toul, France. Courtesy Dwight Gowdey

Bailey Bridge, Kerzel. Corporal Clark, a carpenter, is in the right foreground. Courtesy Earl E. Hall

Timber-trestle bridge, with one pier, or bent, in center, Kerzel. Courtesy Earl E. Hall

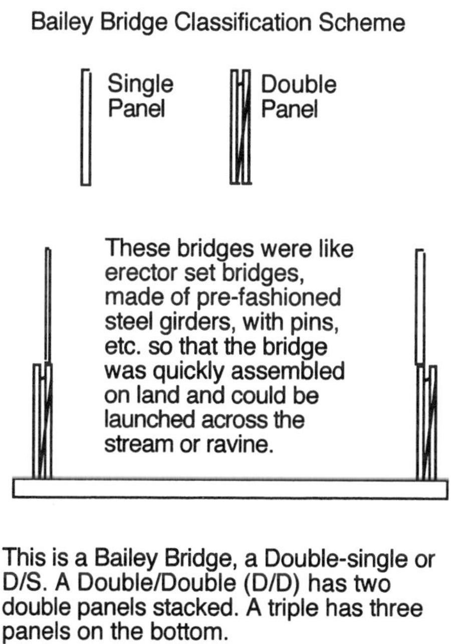

Bailey Bridge classification scheme. Courtesy JCF

Disassembling a Bailey Bridge, Spring 1945. Notice that a black soldier from the attached dump-truck squad is working alongside the men of Company B. Courtesy Earl E. Hall

Ambulance crossing a timber-trestle bridge. Courtesy Earl E. Hall

opener, and spoon, among other things. Each contained breakfast, dinner (now called lunch), and supper (now dinner) packets. K-rations were packaged as a single unit, with a dozen rations packaged in a nailed wooden box for shipment. Dr. Ancel Keys of the Physiological Hygiene Laboratory at the University of Minnesota suggested the contents of this ration, which was then developed by committee.[2]

Notes

CHAPTER 1

1. Casey Deveikis, *The Eager Beaver Regiment: The Regimental History of the 1303 Engineers* (Chicago: 1952). The materials in the National Archives form a very limited collection of reports, plans, photographs, and drawings. The key collection is "Engineering Operations, 1303rd EGSR" (ENRG-1303–0.7), Box 19686, Entry 427, "World War II Operations Reports, 1940–48, Engineers," Record Group (RG) 407 (old RG 94), Records of the Adjutant General's Office, NA, College Park, Md. Some historical reports by the S-3 operations officer are included in the collection too. The unit's "Morning Reports" are at the National Personnel Records Center in St. Louis, Missouri, though a few copies are available in the materials at College Park.

2. The events leading up to the American entry in World War II are multiple and often controversial. There were bitter struggles between isolationists wanting to keep the nation out of war and interventionists seeking to bring it in on the Allied side. The literature is extensive. See, for example, David M. Kennedy, *Freedom from Fear. Part II: The American People in World War II* (New York: Oxford University Press, 1999), chaps. 1–2.

3. Earl E. Hall, e-mail to the editor, 18 July 2003.

4. Easily the best book on the North African campaign is Rick Atkinson, *An Army at Dawn: The War in North Africa, 1942–1943* (New York: Henry Holt, 2002).

5. Earl E. Hall, interviews with the editor, 30 June 2003, 12 Apr. 2004.

CHAPTER 2

1. Dickinson Mounds Museum, "Self-Guided Driving Tour of the Site of Camp Ellis, World War II Army Service Unit Training Camp, Fulton County, Illinois" (Lewiston, Ill., Dec. 1998), http://elaineandmike.us/CampEllis/CampEllisDrivingTour.pdf (11 Feb. 2006), or at http://personal2.stthomas.edu/jcfitzharris/PattonsFightingBridgeBuilders/CampEllisDrivingTour.pdf (courtesy of Dickinson Mounds Museum, 15 Feb. 2006); "Homefront US: Camp Ellis, Illinois," pts. 1–3, World War II Discussion List, http://listserv.buffalo.edu/archives/wwii-1.html (21 Feb. 2006).

2. http://www.elaineandmike.us/CampEllis/CampEllisPostcards.html (10 Apr. 2004).

3. Military time uses a twenty-four-hour clock. For example, 0001 is one minute past

midnight, 0600 is 6:00 A.M., 0930 is 9:30 A.M., 1200 is noon, 1315 is 1:15 A.M., and 2400 is midnight and the end of the calendar day.

4. For a comparative description of general-service and combat-engineer units, see appendix 1.

5. Activated under General Orders (GO) 108, Headquarters (HQ), 6th Service Command, 1 June 43, as announced in GO 25, HQ ASFUTC, Camp Ellis, Ill., 4 July 43, and GO 1, HQ 1303rd EGSR. "History, 1303rd EGSR, 15 May 45–Jan. 46" (ENRG-1303–0.1), Box 19686, Entry 427, "World War II Operations Reports, 1940–48, Engineers," RG 407, Records of the Adjutant General's Office, NA.

6. Lt. Manuel Gonzalez, a Floridian, was the son of a U.S. admiral. He, like the other company commanders, had construction background, owning and operating his own construction firm. He and the other officers were directly commissioned from civilian life and received abbreviated training prior to taking command of their companies. Earl E. Hall, interviews with the editor, 30 June 2003, 12 Apr., 14 June 2004.

Throughout the original diary, officers' names are rendered in capital letters, while enlisted men's names are underscored. Both conventions have been discarded here to enhance the reading experience.

7. Second Lt. Earl E. Hall graduated from Texas A&M in January 1943. Like all ROTC graduates (starting in the summer of 1942), he was given the choice of remaining in the enlisted reserve until called to duty, entering Officer Candidate School (OCS) as a civilian, or enlist as a corporal and go to OCS. Hall enlisted as a corporal and was sent to Camp Polk, where he served as a drill instructor for six weeks until sent to OCS. Afterward, engineer officers went through a six-week engineer-officer course (condensed from nine months) at Fort Belvoir, Virginia. On 1 May 1943 the Division Officer Course was renamed the Cadre Officer Course (COC), taken prior to unit activation. In August the Engineer COC was divided into separate courses for officers in combat-engineer battalions and general-service regiments. Unfortunately, none of the 1303rd's cadre went through the COC. Hall interview, 12 Apr. 2004; Blanche D. Coll, James E. Keith, and Herbert H. Rosenthal, *The Corps of Engineers: Troops and Equipment, Technical Services, the United States Army, World War II* (1958; reprinted, Washington, D.C.: U.S. Army Center for Military History, 1975), 159 [hereafter *CoE, TE*]. After OCS, Hall, Guy Bubb (Company A), and other newly minted second lieutenants were sent to Camp Claiborne, then to Camp Ellis.

8. Training for enlisted men was reduced to eight weeks on 15 March 1942, then gradually expanded to twelve weeks, restoring adequate time for instruction in demolitions, bridging, road construction, and obstacles, all of which were found wanting in the eight-week course. By the summer of 1943, enlisted engineer training was done partly in centralized locations and mostly in the unit. As a result the 1303rd EGSR conducted basic and engineer training for many of its own men. *CoE, TE*, 159–66. As the official history notes, "The assumption that the creation of the engineer soldier could be safely left to his unit had soon proven false" (166).

9. The "cadre" was the core of experienced officers and men who formed and trained a new military unit.

10. First Sgt. Robert Douds was of Irish descent and had heavy-construction experience. Captain Hall had arranged for his commission as a second lieutenant (though with assignment to a different unit), but Douds refused it, preferring to remain a first sergeant and the key enlisted man in the company. Hall interview, 30 June 2003.

11. Howard P. Cooke, Slade, and Murphy were the platoon sergeants of Company B for most of its service.

12. *CoE, TE*, 222–38.

13. Rifle instruction started with "dry firing" the weapon (learning to handle, aim, load, and such without ammunition). Later, using live ammunition, the men fired on targets at several distances (on one-hundred- or two-hundred-yard "known distance" ranges). Firing from four positions (prone, kneeling, sitting, and standing — called "shooting off-hand"), the men were required to "qualify" by putting enough rounds inside the target rings from each distance to achieve the minimum score for "Marksman." More-proficient riflemen earned "Sharpshooter" or "Expert Marksman" qualification.

14. Hall commented that the company received a lot of misfits and the unfit and had to screen these out to get the men with the needed skills. Because other units unloaded unwanted noncommissioned officers (NCOs), the regiment was unable to promote the cadre and other men appropriately until they got rid of the corporals and sergeants who did not have the necessary ability. Because it functioned as something of a replacement depot, the 1303rd was not well trained prior to its departure for Europe. The men had some bridge-building training but were saved by their civilian backgrounds — many were heavy-equipment operators, machinists, carpenters, or welders. All the company commanders had considerable experience in the construction industry, but they received limited training in the business of soldiering. The initial cadre enlisted men all had construction experience and had enough time in service to provide the requisite military knowledge to make the unit work. In some cases men who were physically unqualified were retained because of their skills. For example, a Corporal Hoskins was too old and had a punctured eardrum, but he had the expertise and wanted to serve, so the regiment kept him. Hall interview, 30 June 2003. Hall emphasized that the strength of the 1303rd was in the men of the cadre, who had both prior service experience and construction experience. Hall interview, 14 June 2004.

15. Manpower Training Program no. 5–1, which was the doctrine and schedule in effect when the 1303rd EGSR was formed in the summer of 1943.

16. Dwight Gowdey, "Moving a Bridge 100 Miles," *Military Engineer*, July 1944, 236.

17. Major (later Colonel) Vandenburg was a reserve officer before the war. Hall described him as "a hell of an officer" with civilian construction experience. Lieutenant Colonel Pearson was a graduate of West Point with considerable military service. Hall considered Pearson and Vandenburg critical to the successful training of the regiment. Hall interview, 30 June 2003.

18. Second Lt. Casey Deveikis, the company administrative officer, did not have an engineering background, which probably explains why several officers junior to him by date of commission were promoted to company commands and he was not.

19. During the early months of training, Hall and several other officers took a platoon and worked on a nearby railroad for two weeks, doing whatever needed to be done just to learn the skills. Hall interview, 14 June 2004.

20. Individual training encompassed the knowledge required to be a soldier in the U.S. Army, including rifle training, marching and close-order drill, military courtesies, the Articles of War (the laws governing military service and providing punishment for violations), and physical training. During advanced individual training, a soldier learned the skills of his military specialty, for example, welding or bridge construction. Building upon this, unit training added individual skills and taught the men to work together as a team, on projects ranging from building airstrips to demolitions.

21. Weyman Moseley, Earl Hall, and Lt. Lee A. Ingel of Company A were all fellow cadets at Texas A&M.

22. Another duty assigned to engineer units was clearing mines. Ideally Hall would have returned to Camp Ellis and conducted classes in mine clearing. But since there was never time for this, the men learned in the field, the most demanding form of schooling. Hall coached them until they either got it right or were clearly unsuited for the job. Hall interview, 30 June 2003.

23. The regiment was authorized a certain total of officers, NCOs of each grade, and enlisted men, and within this a particular distribution of skills. Being overstrength meant having more men than authorized, though it could also mean simply having more staff sergeants than allowed. Until the staff sergeants who were unqualified to be in an engineer unit or unfit to be in service at all were removed, the men who deserved the rank could not be promoted.

24. These maneuvers focused on problems in attack and defense, regimental action, and construction under tactical conditions, but they were terminated when deployment orders arrived earlier than expected. "History, 1303rd EGSR, 15 May 45–Jan. 46" (ENRG-1303–0.1), Box 19686, Entry 427, RG 407, NA.

25. An engineer reconnaissance meant going out to look for the materials necessary (timbers, steel beams, and such) to do the various jobs the regiment might be assigned.

26. A "tactical move" means moving as if under combat conditions, with the expectation that enemy forces were nearby and might be encountered.

27. The CCC, Civilian Conservation Corps, was a New Deal agency that recruited young men, organized them into disciplined uniformed work crews, and put them to work on a range of mostly conservation-related projects. Like a number of New Deal relief agencies, the CCC was designed to put money in the hands of the otherwise unemployed as a wage for their labor rather than as a demeaning (in the standards of the time) welfare handout. Most of a corpsman's pay was intended to support his family.

28. Each company was authorized a company commander, three platoon commanders, and an administrative, or executive, officer, normally a captain and four lieutenants. In Company B the lieutenant of the 2nd Platoon became known as the "fifth officer" since each man assigned that command lasted only a short time before moving on for

one reason or another. This constant turnover left the platoon in the hands of its staff sergeant, Howard Cooke, for much of the war.

29. Staff Sergeant Cooke, 2nd Platoon, was too often without a lieutenant, which left him running the platoon. Murphy (1st Platoon), Cooke, and Slade (3rd Platoon) were cadre, and all were very good men and excellent leaders who took their jobs and responsibilities seriously. Slade was a bit older and thus did not quite mesh with the other two platoon sergeants, who were young and, according to Hall, built like football players.

30. POM involved getting the men and the unit ready for deployment. Among other things, this involved issuing the necessary equipment, supplies, and other provisions, all of which had to be inventoried, packed, and readied for shipment.

31. The POE was the departure point for service overseas. The 1303rd staged through Camp Myles Standish near Boston, so Boston was their POE.

CHAPTER 3

1. "History, 1303rd EGSR, 15 May 45–Jan. 46" (ENRG-1303–0.1), Box 19686, Entry 427, RG 407, NA.

2. Ibid.

3. Carlo d'Este, *Decision in Normandy* (New York: Konecky and Kennedy, 1994), 107, 307; Gordon A. Harrison, *Cross-Channel Attack: The United States Army in World War II, The European Theater of Operations* (Washington, D.C.: Center of Military History, 1993), 351.

4. Passage on the *Argentina* was not pleasant. Many men spent the time in their quarters in the holds, seasick and vomiting. Their transit experience to the Pacific would be worlds apart from their trip to Europe. Hall interview, 14 June 2004.

5. This map reference indicates the unit's location, 1.5 miles north of the point designated wL4984 on a particular map of England, "Eng Grid." Following World War I, the British developed a grid system for describing particular geographic points, using a Cassini projection that was later modified for military use over large land sections. This was modeled on the 1915 French adoption of a Lambert map project called Nord de Guerre, for northeast France and Germany, to which was added the Lambert 1, 2, and 3 mappings for western France. The systems used a lowercase and uppercase letter to identify each square in the grid. Points within the grid were identified by two (usually) three-digit codes indicating west-east and south-north orientation inside the square — accurate to about half a kilometer. In Map 6 the figures on the edges of the map are used to derive these codes, and in the legend boxes at the bottom are a description of the methodology. Thierry Arsicaud, "Notes on the Modified British System of Coordinates Used on the European Theatre of Operations during the [*sic*] WWII," http://www.echodelta.net/mbs/eng-welcome.php (accessed 28 Oct. 2005).

6. During World War II the U.S. Army Air Force (USAAF) was one of the three main components of the U.S. Army, the other two being the Army Ground Forces

(AGF), encompassing the combat arms, and the Army Service Forces (ASF), including engineers like the 1303rd EGSR. The USAAF enjoyed a very large degree of autonomy in its operations and planning.

7. The USAAF did not have installations known as airbases overseas but instead had stations — temporary bases. There were so many of these that they were numbered, at first in coherent sequences, but this order broke down under the rapid increase in numbers and their transformation in uses. Station 572 was not an airbase but held ground-support units for the Eighth Air Force. John V. Nicholls, "United States Army Air Force Station Numbers in the United Kingdom in World War Two," Ninth Army Air Force, ETO 1943–1945: The U.S. Ninth Air Force in the United Kingdom, http://www.ninth-airforce.co.uk/stationnumbers.htm (accessed 11 Feb. 2006). Station 541 in Risley was different from Station 572 at Melchbourne Park, a grand estate home taken over by the British government for the duration of the war. The mustard gas "Forward Filling Depot" (FFD) was an American facility, filling sixty-five-pound "LC mustard bombs" for retaliatory use against the Germans. "Site Records: Lords Bridge (FFD4) Forward Filling Depot," Subterranea Britannica, http://www.subbrit.org.uk/sb-sites/sites/1/lords_bridge_ffd/index.shtml (accessed 8 Apr. 2004). Hall notes that the regiment was led into a large field, where the men pitched their tents and stayed until they left for France. The manor house was far in the distance. Hall interview 14 June 2004.

8. This was the single American FFD at Station 572. Five FFDs were built, four by the British, all to the same plan, which included considerable acreage for the storage of incendiary and high-explosive bombs. See "Site Records: Lords Bridge (FFD4) Forward Filling Depot," Subterranea Britannica, http://www.subbrit.org.uk/sb-sites/sites/1/lords_bridge_ffd/index.shtml (accessed 19 June 2004).

9. The communications zone was the logistical support command for, in this case, the United Kingdom. It was divided into three sections, with the 1303rd assigned to the Eastern Base Section. Once on the Continent, the Com Z was the area from the beachhead to the area under control of combat armies — the "Army Zone." Britain then became the rear area.

10. According to the lists of USAAF stations, the one at Langar was actually number 490 in "Notts" (Nottinghamshire) and was a flying station. Station 512 was St. Mawgan, Cornwall, in a different base-section area.

11. The 1303rd EGSR was organized into two battalions of three companies each. Companies A, B, and C were in the 1st Battalion (designated 1/1303), and Companies D, E, and F were in the 2nd Battalion (2/1303). Each battalion had a staff of officers and enlisted men that performed the various personnel (S1), intelligence (S2), operations and training (S3), and supply (S4) functions. Adjutants are charged with maintaining the necessary records, both personnel and institutional; validate all unit general and special orders; and execute many other clerical functions; they also serve as the staff S1 officer. *The Officers' Guide*, 6th ed. (Harrisburg, Penn.: Military Service Publishing, 1941), 252–53.

12. Hall was the only officer trained in mine removal and demolitions. He was sent to every available school but was never given time to train any of the men except for a few in his own platoon — all "on the job." Hall interview, 14 June 2004.

13. A slang term for containers used to remove human waste for disposal.

14. Personnel on "detached service" (DS) were temporarily assigned to duty with another unit or an ad hoc formation. In Hall's case he was assigned to an improvised formation of American engineers and British commandos.

15. The T/O (Table of Organization) sets out the numbers of officers, NCOs, and enlisted men (EM) authorized for the unit, broken down to the company or platoon (or equivalent unit) level. The T/E (Table of Equipment) can be subdivided into Tables of Allowances and Basic Allowances, which are for individual and unit use in garrison (the T/A) and field or combat service (the T/BA)—usually T/E refers to the T/BA. During the war, one often sees TOE (Table of Organization and Equipment) used in place of these multiple terms. *Officers' Guide*, 6th ed., 231–32.

16. The motor convoy sailed on a different ship, and the vehicles, as Hall remembers, were driven off when they disembarked, suggesting an LST (Landing Ship Tank) since all the unit's vehicles arrived together. Hall interview, 12 April 2004.

17. The HMS *Duke of Wellington* was a Liberty ship. The vessel had no nets for disembarkation.

CHAPTER 4

1. Carlo d'Este, *Patton: A Genius for War* (New York: HarperCollins, 1995), 613–25. See also Harrison, *Cross-Channel Attack*, 447.

2. Mark Reardon, *Victory at Mortain: Stopping Hitler's Panzer Counteroffensive* (Lawrence: University Press of Kansas, 2002), 44ff.

3. Alfred M. Beck, Abe Barby, Charles W. Lynch, Lida Majo, and Ralph F. Weld, *The Corps of Engineers: The War against Germany* (Washington, D.C.: Center for Military History, 1985), 393.

4. D'Este, *Decision in Normandy*, 418ff; James Jay Carafano, *After D-Day: Operation Cobra and the Normandy Breakout* (Boulder, Colo.: Lynne Reinner, 2000).

5. *Officers' Guide*, 6th ed., 50–51.

6. The diary does not mention how the regiment got ashore from the *Wellington*. The ship might have tied up to a pier to let the men disembark by stairs, but it is more likely that they went down nets into landing craft and went ashore—except that Hall, as noted earlier, remembers that there were no nets on Liberty ships. The motor convoy came separately, most likely in an LST (Landing Ship Tank), and Hall did not know how the rest of the unit got ashore. Deveikis says the men disembarked from landing craft onto the steel piers built after the invasion, making no mention of how they got from the *Wellington* to the landing craft. *Eager Beaver*, 107.

7. See appendix 4 for a discussion of C- and K-rations.

8. Mess Sergeant Dunbar, a good manager and the best mess sergeant in Hall's experience, ran a superb operation. Invariably, his mess tent had a floor under it so the men would not have to eat in the mud. His mobile kitchen was army issue, but because he was imaginative, a good cook, and scrounged for food, he produced very good meals. He was a buddy of 1st Sergeant Douds and had worked in construction until becoming

a cook in the army. Hall interview, 30 June 2003. Robert S. H. McGlashan claims that Pvt. Wilbur Looper drove the "Diner" and that Privates Woodward, Stuck, Russell, Turner, Richardson, Bailey, Miller, Schoenfeld, Kelly, and Jacobs were the mess crew. McGlashan, "History of Company B," 3 (undated carbon copy [1945], in possession of Earl E. Hall).

9. Company B conducted both "tactical," or "combat," and "road" reconnaissance. Tactical reconnaissance involved going out and looking for Germans, finding out where they were, counting them, and determining how they were armed. The XII Corps often used the company in this fashion beyond American lines. Road reconnaissance involved a careful look at the road network in the area assigned to the unit to determine what needed to be done to make it useable by heavy vehicle traffic. Too often road reconnaissance included the tactical responsibilities as well, for they were to report any Germans they encountered. Engineer reconnaissance—searching for supplies of materials (demolished buildings, iron mills, and such) to be used in the construction of bridges and other structures was done by the battalion S-3 and headquarters company. Sometimes Captain Gonzalez would conduct engineer reconnaissance himself. Hall interviews, 30 June 2003, 12 April 2004.

10. On the road from Cherbourg, tanks, tank destroyers, and heavy trucks continually tore the surface to pieces, as fast as the engineers could put it back together. Their tasks included both "putting the road [material] back when tanks threw it out—with shovels" and "de-mudding" the road, using a grader, or even shovels, to remove mud so that vehicles could gain traction. Regardless of how they did it, the men's mission was to keep the roads open. Hall interview, 30 June 2003.

11. "History, 1303rd EGSR, 15 May 45–Jan. 46" (ENRG-1303–0.1), Box 19686, Entry 427, RG 407, NA.

12. Bradford J. Showedo, *XIX Tactical Air Command and ULTRA: Patton's Force Enhancers in the 1944 Campaign in France*, CADRE Paper 10 (Maxwell AFB, Ala.: Air University Press, 2001), 27–28.

13. Ibid., 30 (map), 32, 45–50.

14. German explosives were generally quite good, and the company routinely used them in the field instead of obtaining U.S. Army issue.

15. "Engineering Operations, 1303rd EGSR" (ENRG-1303–0.7), 27 July–30 Sept. 1944, Box 19686, Entry 427, RG 407, NA.

16. The 764th had General Motors 6×6 dump trucks—six wheels, all powered, hence the "6×6" designation—assigned one per squad of Company B and one platoon per battalion of the 1303rd. There were never enough vehicles. The 764th was an African American unit. First Sergeant Douds put the "colored soldiers" in line for mess as they came in with the platoon rather than segregating them. The sleeping areas were not segregated either, making the company and its attached truck crews, in practice, an integrated unit. Hall interview, 30 June 2003.

17. In this area the regiment deployed for a time as infantry reserves in case the Germans succeeded in cutting the single road the Allies were using and breaking through to the sea. The 1303rd was supported on each side by other units. Fortunately, the Ger-

mans were stopped farther east and the engineers continued their movement through Avranches.

18. Hall was conducting a combat patrol with a squad, examining the roads and looking for Germans, when they came on the causeway to a monastery island. Crossing this, they entered the town and monastery and were told that they were the first Americans to visit since the start of the war. One of the squad, Private Chalom, spoke French; he was quite helpful in establishing communications with locals throughout the European campaign. Hall interviews, 30 June 2003, 14 June 2004.

19. "D/S" refers to the formation of the Bailey Bridge, which could be built with two to four lanes with traffic flow arranged single/single, double/single, double/double, or triple/single. See the appendix 4. See also an exploded diagram at end of "Sir Donald Bailey's Bridge," Rivers Crossed and Their Bridges, 150th Combat Engineer Battalion of WWII, http://www.150th.com/rivers/bailey.htm. "Cl" stands for "class," and the number refers to the load capacity of the span in tons (see chap. 5, note 16 below).

20. The French 2nd Armored Division was attached to the VIII Corps and moved through the area on the roads the 1303rd was maintaining. The French tankers pulled into farmyards, as the crews stopped to visit families and relations, and made messes of the roads as they pulled off and on. They were "a little casual about getting to war." Hall interview, 30 June 2003.

21. The report of items destroyed — unless clearly referring to equipment in the unit's area of operations — are rumors of war and indicative only of their impression of how things were going.

22. Jacqueline was described as a teenage girl, older and "cute as a button," who came into camp and asked several of the men for "McDaniels," a good-looking Tennessean. Obviously they had already met. Hall interview, 14 June 2004.

23. Engineering Operations, 1303rd EGSR.

CHAPTER 5

1. Martin Blumenson, *Breakout and Pursuit* (Washington, D.C.: Office of the Chief of Military History, 1961), 344–50; Martin Blumenson, *Patton: The Man behind the Legend* (New York: Morrow, 1985); d'Este, *Patton*, 411–13. Arguments persist over the reason for the failure to completely close the pocket, with Montgomery often being blamed for slowness.

2. At this point the 1303rd EGSR was the right flank of the XII Corps and thus of the Third Army. But Patton could feel safe entrusting his flank to a general-service unit because Brig. Gen. O. P. Weyland's XIX Tactical Air Command effectively patrolled that area during the advance to the Moselle. Showedo, *XIX Tactical Air Command and ULTRA*, 28, 83ff; Alan F. Wilt, "Coming of Age: XIX TAC'S Roles during the 1944 Dash across France," *Air University Review* 36, no. 3 (1985): 72–76; Lamar Tooke, "Operational Firepower: The Broader Stroke," *Military Review* 81, no. 4 (July–Aug. 2001): 70.

3. Quoted in Showedo, *XIX Tactical Air Command and ULTRA*, 83–84.

4. Ibid., 84. "ULTRA" refers to the breaking of the German "ENIGMA machine"

codes by British intelligence at Bletchley Park. This achievement was so secret that it was not revealed until the 1970s.

5. During this period, Hall and a patrol on a combat reconnaissance entered a town on foot. As they came around a corner, Hall almost ran into a German officer. Both men quickly backed up, turned around, and withdrew. Hall sensed that the Americans were outnumbered and saw no value to alerting hordes of Germans to their presence. Hall interview, 30 June 2003.

6. "Engineering Operations, 1303rd EGSR, 27 July–30 Sept. 1944" (ENRG-1303–0.3), Box 19687, Entry 427, RG 407, NA.

7. Many, if not all, of the 6×6 trucks in the company had a mounting for a .50-caliber machine gun on the roof over the passenger seat, primarily for air defense. Hall interview, 14 June 2004.

8. "History, 1303rd EGSR, 15 May 45–Jan. 46" (ENRG-1303–0.1), Box 19686, Entry 427, RG 407.

9. Major Mylius was known for his messages, which were often irrelevant to the situation at hand.

10. "Engineering Operations, 1303rd EGSR, 27 July–30 Sept. 1944."

11. Showedo, *XIX Tactical Air Command and ULTRA*, 78–116; Blumenson, *Breakout and Pursuit*, 567, 668–69, 686, 689.

12. Blumenson, *Breakout and Pursuit*, 689.

13. Ibid., 666.

14. The headquarters company serviced the battalion and its medical detachment, but its members wanted good meals and got their unit attached to Company B and "Dunbar's Diner" for "rats" (rations) and "qtrs" (quarters or lodging). B probably did not have better quarters but did have a reputation for eating well.

15. Bridge construction over streams, rail cuts, and other interruptions to facilitate a smooth advance was a primary responsibility of the 1303rd. Battalion and regiment S-3 staffs designed the bridges, and the company modified the designs as needed onsite. They often used a Bailey Bridge for the temporary span, then built a wood-and-steel trestle bridge *underneath* the Bailey. Many were built on the rubble of bombed-out bridges, which the engineers regarded as an ideal base. Hall interview, 30 June 2003.

16. Cl 70 (Class 70) one way or Cl 40 (Class 40) two way was a measure of the load capacity of a bridge in tons. Cl 70 refers to seventy tons moving all in one direction (or one vehicle at a time). Cl 40 two way means two-way traffic limited to forty tons in one direction. A Cl 70 one way was equivalent to a Cl 40 two way.

17. A main supply route (MSR) was the road(s) used to move all supplies to a particular unit, whether a division, corps, or army. Even battalions have MSRs.

CHAPTER 6

1. The most direct route, through the outskirts of Paris, is approximately 374 miles. The route the 1303rd took is quite a bit longer. Deveikis estimates that the unit traveled

"some 600 miles from their first bivouac area in Normandy to the Moselle." *Eager Beaver,* 162.

2. This equation-style system of notation tells engineers what kind of bridge (wo, wood; TT, timber trestle; Ka, stone arch; sb, steel beam; etc.) and its dimensions. It also notes the one- and two-way weight-bearing classification (forty or seventy tons). These "equations" are placed in the notes hereafter.

3. Corps engineers maintained an asphalt plant and used gravel or asphalt, as available and appropriate, to repair roadways. Hall interview, 30 June 2003.

4. Men who were hospitalized were dropped from the company roster and listed on the "Detachment of Patients" at their hospital. When patients were discharged, most, certainly infantrymen, were supposed to go to a replacement depot for assignment to a unit that needed men, usually not their original unit. This practice was destructive of unit cohesion and esprit de corps in combat outfits. Somehow, the engineers of the 1303rd were able to return to their regiment, even to their own company, without much problem. Consequently, most men were listed later as returned to duty.

5. Some mills continued operations while others were brought back in service. Hence, new steel beams were often available. Hall interview, 30 June 2003.

6. The 1120th Infantry Regiment (German) was operating against elements of Third Army. On the morning of 27 September 1944, Company F, 134th Infantry Regiment, 35th Infantry Division, was attacked by elements of the 1120th supported by tanks. Butler B. Miltonberger and James A. Huston, "Gremecey Defensive," *134th Infantry Regiment Combat History of World War II,* http://www.coulthart.com/134/chapter_7.htm (accessed, 26 May 2005).

7. Hall remembers bivouacking near Toul one night, and lacking orders for the next day, the company gave a number of men passes into town. Later they received a phone-call order (on a field telephone line) to move out very early the next morning. Hall and a couple of his sergeants went into town to find their missing men. Asking the gendarmes for help, they received a guided tour of the whorehouses. At one, as they went room to room loudly asking for the soldiers by name, there was a shattering of glass; the women reported that three Americans had gone out a window, crashing through the roof of a glass conservatory below. Hall and the sergeants went back to camp empty handed only to learn that their missing men had returned. The unsubstantiated rumor was that they were the three who precipitously left the women's company. Hall interview, 14 June 2004.

8. Rip rapping is a technique designed to protect a slope with a covering material. In this case the protective material chosen was heavy stone, which is used on steeper slopes and where there is a chance of water flowing onto the grade due to either flooding or drainage. Earl E. Hall, telephone interview with the editor, 21 Feb. 2006.

9. The bridge at Toul was a very important crossing point. Construction of the span was hurried because the forward movement of forces across the Moselle River depended on it. Hall interview, 30 June 2003.

10. The several combat engineer battalions assigned to a corps were organized as

a combat engineer group, comparable to a brigade. Sometimes a second group would be attached to a corps, but in the Third Army there were too few combat engineer battalions. The 1301st, 1303rd, and 1306th EGSRs were the only units available, so only one regiment was attached to each corps.

11. Hall used an assault boat to reconnoiter the existing piers and debris piles. The river was running high, and the boat lodged in some trees and tipped over, throwing Hall and his men into the rushing water. Private McDaniels, a Tennesseean who had charmed Mademoiselle Jacqueline in August, was a boatman. Grabbing a small French boat, he rowed out and rescued them all. Hall interviews, 12 Apr., 14 June 2004.

12. A piling was not driven by a big pile driver but with either a small unit, a crane and a large weight, or by hand. Hall interview, 30 June 2003.

13. Engineering drawings were done by the staff of the regimental — for major projects — or battalion — for smaller projects — S-3 (plans and operations or engineer officer). The platoon then modified these to fit the materials and equipment available. Plans were based on "engineer reconnaissance" by company or platoon patrols rather than on surveys. Designs were modified on the ground as needed, as this entry indicates. Hall interview, 24 Apr. 2004.

14. Hall recounts having General Patton appear, chromed siren, MP outriders, and all. Patton informed him that the bridge would be finished at a particular time and day. Hall, knowing it would not be done by that time, informed the general that a certain sequence had to be followed and that construction would be completed by noon of the following day. Patton glowered and, apparently assuming that any lieutenant who would correct the commanding general (just below God in the pantheon of superiors) probably knew what he was doing, left without further comment. When the bridge was finished as forecast, the general sent the engineers an inscribed photograph. Hall interview, 30 June 2003.

15. A drag line is a bucket on a cable attached to a crane. The crane "throws out" the bucket and then drags it back, filled with material. Hall interview, 12 Apr. 2004. The 840th Engineer Aviation Battalion was working on an airfield at Toul/Crois de Metz at this time. "ETO Airfields: General Construction Information," IX Engineer Command, http://www.ixengineercommand.com/airfields/general.php (accessed 21 Feb. 2006.)

16. This mixer was quite likely "borrowed" from a civilian firm. Whether the company knew of this "requisition" of equipment or not is unknown.

17. "Engineering Operations — Nov. 1944," 1, "Engineering Operations, 1303rd EGSR, 27 July–30 Sept. 1944" (ENRG-1303–0.3), Box 19687, Entry 427, RG 407, NA.

18. A "tactical" bridge is one serving an immediate combat need. Often a ponton or Bailey Bridge, it would normally be built under fire by combat engineers. Hall interview, 30 June 2003.

19. "Engineering Operations — Nov. 1944," 19–20.

20. Ibid., 1–2.

21. The 61st Article of War covered absences without leave (this now falls under Article 86 of the Uniform Code of Military Justice). Article 96 was the "general" article

(now Article 134, UCMJ) covering unbecoming conduct and such. Dr. Donald Bittner, Marine Command and Staff College, was most helpful on this issue.

22. "Engineering Operations—Dec. 1944," 2, "Engineering Operations, 1303rd EGSR, 27 July–30 Sept. 1944" (ENRG-1303–0.3), Box 19687, Entry 427, RG 407.

23. Huntington was a lawyer from Massachusetts with political connections, later given a direct commission as a second lieutenant.

24. Hughes was given to exaggerating his nonexistent heroic combat exploits in letters home. Hall interview, 12 Apr. 2004.

25. Since company clerks worked in the company headquarters, not regimental or battalion, this entry is somewhat confusing; no light can be shed on what is really meant here. It may indicate that regimental headquarters needed extra clerks to handle materials relating to the company, and Brehl and Groom were simply assisting regimental clerks with the paperwork.

26. In both France and Luxembourg, the company recruited civilian road workers to help maintain the roads. While these workers were not paid by the unit (and probably not by the army), they often received rations while working. Other civilians were recruited to assist in clearing and rebuilding work, with rations as their pay.

27. The Third Army's movement to Bastogne and Luxembourg was not done in the orderly manner prescribed by staff colleges. The 1303rd received oral movement orders regarding the route they would take and the precise times they were to enter each intersection and when the next convoy would enter the same intersection. There were too few military policemen to control movement. Captain Hall said the unit drove at their best speed to make their intersection on time—they could not wait for an opening if they did not make their slot and had to be clear because the crossing unit would not stop. Hall interviews, 30 June 2003, 14 June 2004.

CHAPTER 7

1. Williamson Murray and Allan R. Millett, *A War to be Won: Fighting the Second World War* (Cambridge, Mass.: Harvard University Press, Belknap Press, 2000), 466–71.

2. Two useful sources on this campaign are Hugh M. Cole, *The Ardennes: Battle of the Bulge, the United States Army in World War II* (Washington, D.C.: Office of the Chief of Military History, 1965); and Max Hastings, *Armageddon: The Battle for Germany, 1944–1945* (New York: Knopf, 2004). See also d'Este, *Patton*, 676ff.

3. Hall interviews, 30 June 2003, 14 June 2004. Geoffrey Perret notes that in 1941 or 1942 the operational conditions—icy roads and deep snow off road—would have made Patton's northward shift an impossible mission. *There's a War to be Won: The United States Army in World War II* (New York: Random House, 1991), 411.

4. This number is quite likely exaggerated. Even if the Luftwaffe had enough fighter aircraft available to devote to this operation, the diarist would have had to spend his time counting the enemy planes in the air—a most improbable event in itself. Nevertheless,

this entry indicates that there were more German aircraft in action than "normal," probably by a factor of three, four, or even five.

5. "Engineering Operations — Dec. 1944," 20, "Engineering Operations, 1303rd EGSR, 27 July–30 Sept. 1944" (ENRG-1303–0.3), Box 19687, Entry 427, RG 407, NA; Hall interview, 14 June 2004.

6. This card is reproduced in Deveikis, *Eager Beaver*, 229. Patton's greetings read: "To each officer and soldier in the Third United States Army, I wish a Merry Christmas. I have full confidence in your courage, devotion to duty, and skill in battle. We march in our might to complete victory. May God's blessing rest upon each of you on this Christmas Day." The prayer on the other side reads: "Almighty and most merciful Father, we humbly beseech thee, of Thy great goodness, to restrain these immoderate rains with which we have had to contend. Grant us fair weather for Battle. Graciously hearken to us as soldiers who call upon Thee that, armed with Thy power, we may advance from victory to victory, and crush the oppression and wickedness of our enemies, and establish Thy justice among men and nations. Amen."

7. It was reliably reported that Huntington, who was earlier reduced to private from corporal at his own request, received his commission due to the intervention of a large Boston financial institution and orders from Washington. Hall notes that Huntington was a good man. Stoudt, a newspaperman in civilian life, earned his lieutenancy, and 1st Sergeant Douds deserved a commission but refused it. Hall interview, 30 June 2003.

8. Neither Captain Hall nor other World War II veterans could identify what "FPL" means. The editor's best guess is "final protection line." See ["F"], "Military Acronyms, Initialisms, and Abbreviations," Military Lexicon, Federation of American Scientists, http://www.fas.org/news/reference/lexicon/acf.htm (accessed, 1 June 2005).

9. Chalom said that the boy claimed he was an orphan. The men of the company fed and took care of him, and he accompanied them as a kind of mascot. Wilbur Looper was particularly attached to the boy. "Butch" died a day or so later. Hall interviews, 30 June 2003, 14 June 2004. Other units also adopted children who, frequently uniformed and armed as U.S. soldiers in miniature, then accompanied the Americans.

10. The diarist again provides specific (and large) numbers regarding German losses. The figures themselves should be ignored, the relevant point being that the engineers thought or knew the enemy was being beaten.

11. One of the motor sergeants was especially good at procuring supplies, particularly lumber and vehicles. Reportedly, in civilian life he had been the head of a car-theft ring in Detroit, taking his loot across the Canadian border for liquidation. Hall interview, 14 June 2004.

12. S-3 (plans and operations staff officer) in an engineer regiment developed the construction design, tracked down the necessary materials, and allocated the work. Maj. John W. Polick was a good officer, younger (not a "crusty construction man"), who valued and recognized good work and good people. Hall interview, 14 June 2004.

13. Earl Hall does not remember the source(s) of this kind of information, and it is quite possible that the information came through quasi-official channels.

14. Douds was awarded the Bronze Star in GO 7, HQ 3rd Army. "History, 1303rd EGSR, 15 May 45–Jan. 46" (ENRG-1303–0.1), Box 19686, Entry 427, RG 407, NA.

15. Hall and several men arrived in Mersch in time to cross the bridge over the ravine in the center of town and observe, at the bottom of the gorge, an American soldier making the intimate acquaintance of a young woman. While the two were in a grove of trees and thus invisible from ground level, they were quite visible to anyone on the bridge above. Hall interview, 30 June 2003.

16. In France and Luxembourg, roads and highways were numbered and differentiated in quality and function by the following letters: N, Route Nationale, comparable to U.S. national highways; D, Chemin Départmental, comparable to American state highways; GC (or Gc), Chemin de Grande Communication, apparently comparable to American county highways; IC (or Ic), Chemin de Intérêt de Communication, which seem comparable to low-grade county roads, though not a township road; and VO, Chemim Vincial Ordinaire, apparently comparable to township roads in America. From the legend on Sheet 54 (Cherbourg–Rouen, France), European Road Map (scale 1:200,000), Geographical Section, General Staff 4238 (War Office, 1941), U.S. Army Map Service, 1942, N4859–E104/51×256W, owned by and used with permission of Doug Bekke (14 June 2005).

17. Major Mylius was relieved of battalion command when the strain of command in a combat zone became too much for him. He returned to duty in the zone of the interior. Captain Gonzales became battalion commander. Hall interviews, 30 June 2003, 14 June 2004.

18. "Engineering Operations — Feb. 1945," 1–2, "Engineering Operations, 1303rd EGSR, 27 July–30 Sept. 1944" (ENRG-1303–0.3), Box 19687, Entry 427, RG 407.

19. The Diekirch bridge was classified in the diary

$$\frac{sb\ w20\ L180'\ H20'}{2wc140\ 1wc170}$$

a steel-beam span 20 feet wide, 180 feet long, and 20 feet high, with two-way traffic capacity of forty tons, one-way, seventy tons.

20. The company was exempted from this draft and lost no men to the infantry. Hall interview, 30 June 2003. Given the engineers' combat role, this was rather reasonable.

21. When a tie line broke loose, Stoudt was not supposed to be where he was — he was no longer a sergeant — and sustained serious injuries. Sergeant Cooke took him to the nearest hospital without pausing for permission while the men carried on. Stoudt was arguably the best of the many lieutenants assigned to the 2nd Platoon, and his loss was keenly felt, not least by Cooke, who again took charge of the platoon. Hall interview, 14 June 2004.

22. This advance by the XXI Army Group (Operation Veritable) and the First and Third Armies (Operation Grenade), one of some significance, hardly figured in the 1303rd's operations. The engineers still had bridges to repair or build, roads to maintain, and their other routine duties to perform. Their own push east would come later.

23. By order of Headquarters, European Theater of Operations, on 17 March 1945,

2nd Lt. Logan H. Stoudt was awarded the Legion of Merit for his performance of duty. "History of the 1303rd EGSR, 1943–45," (ENRG-1303–0.7), Box 19686, Entry 427, RG 407, NA.

24. The Christnach bridge was classified in the diary

$$\frac{\textit{wo 1w14' L32'}}{\text{Cl 6}}$$

a one-way wooden bridge fourteen feet wide and thirty-two feet long.

25. The Lintgen bridge was classified in the diary

$$\frac{\textit{wo 1w12' L46'}}{\text{Cl 7}}$$

a one-way wooden bridge twelve feet wide and forty-six feet long.

26. The civilian workers, mostly the same people who had maintained the roads in Luxembourg during peacetime, were not paid for their work, though they did receive food rations on the job. Hall interview, 14 June 2004.

27. Known as "club mobiles," these ungainly doughnut bakeries moved about not far behind the lines. Each vehicle and crew served the same units and became well known to the men.

28. These reports of the air war are based on whatever news became available from higher headquarters, the "rumor mill," and possibly observation of many flights involving large numbers of aircraft. The actual number of sorties flown or aircraft involved were quite beyond the diarist's knowledge, as was the fate of Dresden or Chemnitz. This illustrates the effectiveness of the information-transfer mechanisms operating within the Third Army or most any army. Somehow the news circulates, changing mightily as it flows.

29. Fast was the blacksmith, "a big bulking guy." While in Germany, both Captain Hall and 1st Sergeant Douds had offices. Fast charged into Douds's office one day saying, "I'm going to get that guy Hall and kill him." Douds got up, hit Fast in the chin, putting him out—through the door. Hall interview, 30 June 2003.

30. Many, if not most, combat units earned poor inspections ratings. Often the best units fared the worst in confrontations with the inspectors general, and the best officers received the poorest ratings.

31. The Christnach bridge was classified in the diary

$$\frac{\textit{TT 2w9 48'L}}{\text{Cl 20}}$$

a two-lane timber-trestle span forty-eight feet long, each lane nine feet wide.

32. S.Sgt. Howard P. Cooke, Sgt. Robert L. Pennington, and Pfc. John J. Laffery were awarded the Bronze Star in GO 49, 3rd Army. Pfc. Michael NMI [No Middle Initial] Dziewnka was awarded the Bronze Star in GO 50, 3rd Army. "History, 1303rd EGSR, 15 May 45–Jan. 46."

33. The usual problems of knowledge exist here. The men of Company B likely learned, quickly since it was major news, of the Remagen bridgehead. Its importance was clear to engineers, and the fate of those who failed was clear to anyone who had any experience with the Nazi mentality.

CHAPTER 8

1. Andreé became another "mascot" for the company, somewhat like "Butch" had been, but she had a family and did not travel with the Americans into Germany. Hall interview, 30 June 2003.

2. The bridge was classified in the diary

Sb 2w10 L38'

2wc140 1wc170

a steel-beam span thirty-eight feet long, with two lanes each ten feet wide.

3. Murray and Millett, *War to Be Won*, 477–79; d'Este, *Patton*, 710ff; Perret, *There's a War to be Won*, 436ff.

4. The data on German losses would not be available to the regiment. This is rumor and invention.

5. The battalion and regiment selected the men for leave. Hall interview, 30 June 2003.

6. In the Third Army Engineer Section report of Brig. Gen. John F. Conklin, he notes that LCMs (Landing Craft, Medium) and LCVPs (Landing Craft, Vehicle, Personnel) were used at a bridge at Trier on the Moselle. Photographs of the various crossings suggest either assault boats or LCMs and LCVPs were used but not larger vessels. Getting an LSI (Landing Ship, Infantry) overland would have been a daunting task since such vessels are over 120 feet long. Deveikis notes that the 2nd Battalion had forty-two men trained in the operation of LCVPs who were on detached service to assist in the Rhine crossing. *Eager Beaver*, 289.

7. Murray and Millett, *War to be Won*, 479–80.

8. Attacks by German ME 262 jets were common at this time. Hall remembers being strafed by one, and despite his best efforts, he could only follow it with his .50-caliber machine gun, the tracer stream curving far behind the jet. The aircraft was simply too fast for gunners to properly lead and hit.

9. The bridge was classified in the diary

Sb 2w10 L125'

2wc140 1wc170

a steel-beam span 125 feet long, with two lanes each 10 feet wide.

10. The 1303rd EGSR was *assigned* to the Third Army and *attached* to the XII Corps; at this point the unit was finally *assigned* to the XII Corps. "Assignment" means "ownership," while "attachment" means use subject to the needs of the higher organization that "owns" the unit in question.

11. These SS soldiers were members of the Nazi regime's Schutzstaffel (secret police and private army), most likely part of the "Waffen SS," the organization's military arm. Another part of the SS ran the concentration camps, and yet a third the killing squads (Einsatzgruppen) that roamed the eastern front, killing Jews, partisans, and other undesirables.

12. Lieutenant Moseley's mother was a WAC [a member of the Women's Army Corps] at some higher headquarters.

13. Deveikis notes that a detail from the 1303rd was sent to the salt mines at Merkers, where gold, stolen art works and statuary, and other valuables were hidden by the Nazi regime, to reestablish light and power. *Eager Beaver*, 320.

14. "Engineering Operations—Apr. 1945," 6, "Engineering Operations, 1303rd EGSR, 27 July–30 Sept. 1944" (ENRG-1303–0.3), Box 19687, Entry 427, RG 407, NA.

15. The Schluchtern bridge was classified in the diary

$$\frac{sb\ 2w10\ L65'}{2wc140\ 1wc170}$$

a steel-beam span sixty-five feet long, with two lanes each ten feet wide.

16. The Kerzell bridge was classified in the diary

$$\frac{sb\ 2w10\ L50'}{2wc140\ 1wc170}$$

a steel-beam span fifty feet long, with two lanes each ten feet wide.

17. "Silver Taps," a rendition played by three or more trumpeters, in harmony, was a Texas A&M tradition.

18. The Harras bridge was classified in the diary

$$\frac{sb\ 2w10\ L38'}{2wc140\ 1wc170}$$

a thirty-eight-foot-long steel-beam span, with two lanes each ten feet wide.

19. Earl Hall reported entering Bayreuth on reconnaissance. He found a tanker about to blow the door of the opera house. Hall was looking for quarters, and this was one of the few intact buildings available. Ranking the tank commander, who did not want to leave any possible enemy behind him, Hall ordered them away and took possession of the site. As a result, Company B was quartered in the opera house, and Hall used Madam Wagner's suite as his quarters and company command post. A two-person USO troop of young ladies, whom Hall remembers as being "cute as a button," performed for the unit on the world-famous stage. In the shrine of Wagnerian music, the women sang "Don't Fence Me In!" This was the first USO performance the men of Company B enjoyed overseas.

20. The Bayreuth bridge was reclassified in the diary

$$\frac{sg\ 2w12\ L120'}{2wc140\ 1wc170}$$

a steel-girder span 120 feet long, with two lanes each 12 feet wide.

21. Richard Reese was the special-services sergeant, responsible for what is now called morale, welfare, and recreation programs. Some months after he was selected for promotion, Reese objected to an evaluation by the company commander and was on the verge of insubordination until the threat of a general court-martial brought a return of manners. He later transferred to another company. Hall interview, 14 June 2004.

22. All travel was based on priorities. With equal priorities, rank determined who got the first available seat. Neither Deveikis nor the detachment had enough priority or rank to get a flight, and obtaining land transportation was almost as difficult.

23. The Schwarzach bridge was classified in the diary

sb 2w10 L95'

2wc140 1wc170

a steel-beam span ninety-five feet long, with two lanes each ten feet wide.

24. Earl Hall and Guy Bubb went through OCS, Camp Claiborne, and then to Camp Ellis as cadre together.

25. The Rottenegg bridge was classified in the diary

sb 2w10 L90'

2wc140 1wc170

a steel-beam span ninety feet long, with two lanes each ten feet wide.

26. Hall claims that the commander of the 12th Panzer Regiment surrendered to him in Passau, then asked when his men would be given back their arms so that, with the Americans, British, and French, they could turn east and attack the communist hordes. He was quite shocked to learn that no such effort was planned, that his men were to be held as PWs. Hall interview, 30 June 2003.

27. Major Gonzalez, the battalion commander, was from Florida and capable of operating a power boat on the Danube or the Atlantic Ocean. Hall interview, 30 June 2003.

28. The XII Corps commended the 1303rd EGSR for its work in support of the corps. The three EGSRs sent to the Pacific were chosen because they were indeed good units. *CoE, TE*, 306.

29. Also reported in "History, 1303rd EGSR, 15 May 45–Jan. 46" (ENRG-1303–0.1), Box 19686, Entry 427, RG 407, NA.

CHAPTER 9

1. One of the more interesting works on the war against Japan is Richard B. Frank, *Downfall: The End of the Imperial Japanese Empire* (New York: Random House, 1999). Frank draws heavily upon the MAGIC and other intelligence intercepts in writing his history. In two chapters, 9 and 12, he thoroughly discusses the assumptions and experiences that supported the casualty estimates for Operation Downfall. The 1303rd was intended to support the 1st Marine Division in landings on Kyushu in November 1945 as part of the initial invasion forces.

2. Quoted in Murray and Millett, *War to be Won*, 419.

3. The captain of the USS *General John Pope* used his public-address system as the ship entered the canal to urge the workers to speed the passage as he had high-priority troops for the Pacific onboard. Hall interview, 30 June 2003. This voyage was much more enjoyable than the company's first Atlantic crossing on the *Argentina*. The *General John Pope* was designed to be a transport. Onboard the unit officers had no responsibility for their commands; the ship's complement of marines handled things. Ibid., 14 June 2004.

4. Earl Hall played a lot of poker during the crossing and sent his mother a substantial amount of money on arrival in the Philippines. Ibid., 14 June 2004.

5. The acronym "TAT" means "to accompany troops." ["T"], "Military Acronyms, Initialisms, and Abbreviations," Military Lexicon, Federation of American Scientists, http://www.fas.org/news/reference/lexicon/act.htm (accessed, 7 June 2005).

6. Hall thinks that even this entry makes more of the episode than is warranted. 1st Sergeant Douds did put the absent men on sick call to avoid trouble, and the incident was worked out rather quickly. Hall interview, 30 June 2003.

7. The regiment had almost no equipment assigned to it at this point. Needing trucks, they acquired 6×6s from the air force. Onboard the LST someone recognized one of the trucks as USAAF property and reported it. Hall almost received a general court-martial as a result. He notes that in Europe, if someone needed something that was not being used by another unit, he took it, marked it as belonging to his unit, and went on about his business. If caught, he simply gave the equipment back and went looking for more.

8. John W. Dower, *Embracing Defeat: Japan in the Wake of World War II* (New York: W. W. Norton, 1999), 38, 43–63.

9. Ibid., 61, 130.

10. "History, 1303rd EGSR, 15 May 45–Jan. 46" (ENRG-1303–0.1), Box 19686, Entry 427, RG 407, NA. Deveikis discusses the 1303rd's deployment in Japan, though without much detail on Company B. See *Eager Beaver*, 370–97.

11. "History, 1303rd EGSR, 15 May 45–Jan. 46."

12. There is some evidence that Lieutenant Ryan was perceived as making himself the star of this show, to the annoyance of at least the keeper of the diary.

13. "History, 1303rd EGSR, 15 May 45–Jan. 46."

APPENDIX 1

1. *Officers' Guide*, 6th ed., 15.

2. *CoE, TE*, 221 ff.

3. Ibid., 238–39.

4. Ibid., 222, 234ff.

5. Ibid., 222ff.

APPENDIX 2

1. *Officers' Guide*, 6th ed., 165ff.

APPENDIX 3

1. *CoE, TE*, 159–74; "History, 1303rd EGSR, 15 May 45–Jan. 46" (ENRG-1303–0.1), Box 19686, Entry 427, RG 407, NA.

2. *CoE, TE*, 299.

3. Ibid., 300.

4. Ibid., 201–203.

5. Ibid., 305–306.

6. "History, 1303rd EGSR, 15 May 45–Jan. 46."

APPENDIX 4

1. Hall interview, 14 June 2004.

2. Franz A. Koehler, "Army Operational Rations — Historical Background," chap. 1 of *Special Rations for the Armed Forces, 1946–53*, QMC Historical Studies, Series 2, No. 6 (Washington, D.C.: Historical Branch, Office of the Quartermaster General, 1958). http://www.qmfound.com/army_rations_historical_background.htm#The%20K%20Ration (accessed, 8 Apr. 2004).

Index

In the absence of most first names, rank is used for identification. Still, there are occurrences of two or three men with the same last name, either with no rank listed or of the same rank. Reader should not assume these refer to a single man; separate entries are shown.

Numbers in *italics* refer to illustrations.

ISBN 978-1-58544-550-9
52995
9 781585 445509

This quaint little town, tucked in the heart of Pennsylvania Dutch country, was founded in 1754. First called Cross Keys after an old tavern that stood at the crossroad, the name evolved into Intercourse by 1814. The name itself inspires blushes, but has also contributed to a boom trade in this tucked away town. Visitors to the Lancaster area and Amish country flock here, and a cottage industry has sprung up to serve the curious hordes. Stores overflow with folk art indigenous to the area, and restaurants serve up heaping portions of local cuisine.

Apparently, way back in 1841, George Brungard foresaw a grand and busy future for the crossroads, then home to five buildings. Around that time he purchased 48 acres, divided them into 151 lots, and asked $250 apiece for them. Takers, it seems, were slow to buy them.

The village enjoyed slow and steady growth through the decades. A second tavern, Travelers Rest Inn, was built there in 1827, and by 1883 there were 54 dwellings, 280 people, two stores, a feed store, a blacksmith, a wheelwright, two harness makers, three carpenter shops, and a shoe store, butcher shop, and cabinet shop. There were also three physicians, a dentist, and one tobacco warehouse. Today the population is estimated at about 900 persons, though the area is not an official town.

A stagecoach was established to make the run between Intercourse and Lancaster in 1898, and in 1909 the town got streetlights and telephone/telegraph service. A year later a motor truck replaced the stagecoach, and in 1916, the town got its first gas pump.

The town has stood conservative against attempts to poke fun at its name. In 1966, they turned a pornography company, Eros Publishers, away when it attempted to establish a mailing address in Intercourse. More recently, an attempt to stage a rock concert event in the area was turned down.

It's a love/hate relationship between conservative members of the community and those who enjoy the crowds who travel to this remote village to say that they've "been to Intercourse." In 1956 there were petitions circulated in favor of calling the village Cross Keys again, or Smokerville, but these were unsuccessful. For better or worse, the town sticks by its name, while puzzling over its origin. Three theories exist, none proven. One line of thought is that the name relates to a racetrack once located east of the village. The entrance at the east of town was referred to as "enter course" and gradually morphed into the community's name. Another theory holds that the intersection or "intercourse" of the two roads that cross in the village — Old King's Highway (now Old Philadelphia Pike) which ran east to west and The Newport Road running south — spawned the name. Finally, in writing dating to the early days of the town, the word "intercourse" was commonly used to refer to casual social interaction and friendship, thus a lovely name for a small village.

Kitchen Kettle Village is testament to the hordes that arrive and enjoy Intercourse year round. An enormous parking lot serves buses, cars, and buggies alike. Inside the complex, a series of wonderful shops provide crafts, canned goods, and fresh vittles to visitors. A band provides country music on clement days.

DAS
CAFE HAUS
HomeMade
Lite Fare • Bakery
Coffee
Breakfast • Lunch • Dinner
ADT

BAKE SHOP
KITCHEN KETTLE

"the quilt of your dreams"
Village
Quilts
Kitchen Kettle Village
Intercourse, Pa.

OLD-FASHIONED SUGARY POPCORN
KETTLE KORN
KETTLE KORN
TRY SOME NOW!!... TAKE A LARGE ONE HOME!!
WWW.PAPPYSKORN.CO

KITCHEN KETTLE
Intercourse, Pennsylvania

Jams & Jellies
Relishes
Create your own sampler
Flavors available in mini jars:

Most of the historic buildings in Intercourse have been converted to commerce. A steady stream of foot traffic makes them prime targets as purveyors of quilts and other handicrafts. A quilt museum and an astonishing selection of quilting fabrics for sale make this a mecca for quilters and crafters. Customers and the merely curious arrive via car, caravan, and carriage.

ZIMMERMANS
MEATS
Phone

The Old Country Store
Intercourse, PA.
Quilts, Fabrics, Country Crafts & Wares
Operated by The People's Place
The Old Country Store
The People's Place
QUILT MUSEUM
Antique Amish Crib Quilts from the Midwest
– AND –
Antique Mennonite Quilt Masterpieces from Ontario
"A 15th Anniversary Exhibit"
• Antique quilts
• Exquisite museum setting
• Very informative
HERITAGE
3510

WOOLRICH
The Original Outdoor Clothing Company
Village Outfitters
THE INN
LODGING INFORMATION

The People's Place
QUILT MUSEUM
Antique Amish Crib Quilts
from the Midwest
– AND –
Antique Mennonite Quilt
Masterpieces from Ontario
"A 15th Anniversary Exhibit"
March 21 through October 31, 2003
• Antique quilts
• Exquisite museum setting
• Very informative
VILLAGE POTTERY
OPEN
HERITAGE
OPEN

OLD
CANDLE
BARN

STAGECOACH
SHOPS
of Intercourse

Lapp's Furniture & Crafts
Rockers · Hope Chests
Jewelry Chests · Mailboxes
Large Display Showroom
WAGONS FOR SALE
PINE FURNITURE

VILLAGE PEDDLAR
INTERCOURSE
BIRD

Amishland Prints
ARTIST Xtian Newswanger
www.amishlandprints.com
3504
OPEN

THE
AMERICAN MILITARY
EDGED
WEAPONRY
MUSEUM
STOP
EAST
340
OPEN

PEDDLERS
COTTAGE
Crafts
Furniture
Antiques
Apparel

PEDDLERS
COTTAGE
Crafts
Furniture
Antiques
Apparel

Fabrics
Quilts
Crafts
ETC.
HANDQUILTED QUILTS
WOODEN ACCENTS IN OAK & PINE
HANDMADE LAWN ITEMS
Woven Afghans
Appliqued Sweatshirts
PARKING
PARKING
3466 OLD PHILA. PIKE

ICE CREAM
COLD DRINKS
SODAS & WATER
3582

PRIVATE PARKING
VIOLATORS WILL BE TOWED AWAY AT OWNERS EXPENSE

THE
KICK N Stitch
BROOM
SHOP
&
MUSEUM
AN AUTHENTIC 1800's WORKING SHOP

JAMS & RELISHES

ONLY
THIS SIDE OF STREET
ouse-Cafe
DAS CAFE HAUS
HomeMade
Lite Fare · Bakery
Coffee
Breakfast · Lunch
Coca-Cola

EBY'S HOUSE
WELCOME
Bless Our Home
Welcome to Our Garden
Welcome Friends

rseTradingPost
OPEN
WELCOME

FLOWER SHOP
Welcome

OPEN

Intercourse

The LUNCH BASKET
SUBS SOUPS
ICE CREAM
BREAKFAST
THIS SIDE OF STREET
77

Intercourse Village
Bed & Breakfast
Suites

Welcome
· to ·
Our
Kitchen

3448
Kings
Guest House
717-768-3425

THAT
PICTURE
PLACE
Prints, Posters
& More

ROCKING HORSES
FOR SALE
Many sizes are available
LAPP'S COACH SHOP
INTERCOURSE, PA.
PARKING
IN
REAR
CLOSED
SUNDAYS

Two worlds collide, and tourists come to gawk at the unique culture of the Amish. The gawking goes both ways in Intercourse. For the young Amish, the “English” in their outrageous attire are equally curious. For the elder Amish,

the snapping cameras and stream of questions often are an unwelcome nuisance. The Amish still conduct day-to-day business in the town, and their immaculate farms surround the tiny hamlet.

RICE

Intercourse Village
Bed & Breakfast
Suites
1909
VACANCY
Getty
Food Mart

Intercourse is surrounded by farmland. Its first and foremost purpose as a commercial center is serving the local farm community. Equine watering holes, tractor sales, and buggy shops cater to the farmers. For tourists, the town's rural nature is one of its foremost delights.